SpringerBriefs in Electrical and Computer Engineering

More information about this series at http://www.springer.com/series/10059

Wanrong Tang • Ying Jun (Angela) Zhang

Optimal Charging Control of Electric Vehicles in Smart Grids

 Springer

Wanrong Tang
Department of Information Engineering
The Chinese University of Hong Kong
Shatin, Hong Kong

Ying Jun (Angela) Zhang
Department of Information Engineering
The Chinese University of Hong Kong
Shatin, Hong Kong

ISSN 2191-8112   ISSN 2191-8120   (electronic)
SpringerBriefs in Electrical and Computer Engineering
ISBN 978-3-319-45861-8   ISBN 978-3-319-45862-5   (eBook)
DOI 10.1007/978-3-319-45862-5

Library of Congress Control Number: 2016950032

Printed on acid-free paper

This Springer imprint is published by Springer Nature
The registered company is Springer International Publishing AG
The registered company address is: Gewerbestrasse 11, 6330 Cham, Switzerland

Preface

.

The wide penetration of renewable energy sources and plug-in electric vehicles (PEVs) has imposed significant challenges to the design and operation of the power grid. In particular, the increase in the intermittent renewable sources, such as solar and wind power, seriously affects the provision of system services that balance supply and demand. Such services include frequency regulation, voltage control, and the control and management in day-ahead, hour-ahead, and real-time operation. Utilizing the energy storage system (ESS) in power grids is considered an effective mechanism for absorbing the fluctuation of energy generation and consumption. Besides traditional ESSs, such as pump hydro, the increasing number of PEVs can be viewed as an emerging type of battery energy storage systems (BESSs) that are widely available at the distribution level. This book studies the optimal online charging control of BESS and PEVs, with the aim to absorb the random fluctuation in the power supply as well as demand and reduce the additional burden on the grid due to massive EV penetration. Both the theoretic analysis and numerical results show the effectiveness and efficiency of the proposed online control techniques.

This book not only provides researchers with the latest research results timely and extensively but also presents a comprehensive overview of the online charging control techniques. In particular, the online control techniques have strong practicability since they do not rely on any noncausal knowledge of future information. The researchers, operators of power grid, and EV users will find this to be an exceptional resource.

Shatin, Hong Kong   Wanrong Tang
June 8, 2016   Ying Jun (Angela) Zhang

Acknowledgements

This book becomes a reality with the kind support and help of many individuals in the past few years. We would like to extend our sincere thanks to all of them. The work described in this book was supported in part by the General Research Funding (Project number 14200315) from the Research Grants Council of Hong Kong and Theme-Based Research Scheme (Project number T23-407/13-N).

Contents

Acronyms

AVR	Average Rate
BESS	Battery Energy Storage System
CAES	Compressed Air Energy Storage
ESS	Energy Storage System
EV	Electric Vehicle
FESS	Flywheel Energy Storage System
MPC	Model Predictive Control
OA	Optimal Available
ORCHARD	Online cooRdinated CHARging Decision
PEV	Plug-in Electric Vehicle
PHS	Pumped Hydro Electric System
SAA	Sample Average Approximation
UPS	Uninterruptible Power Supply
VRB	Vanadium Redox Flow Battery
VSS	Value of the Stochastic Solution
YDS	Yao, Demers and Shenker

Chapter 1
Introduction

1.1 Motivations

Renewable energy sources have been widely adopted to reduce carbon dioxide emissions and the dependence on fossil fuels. The large-scale integration of renewable energy sources imposes significant challenges to future power grids, mainly because the power generation from renewable energy sources is intermittent and fluctuating. Uncertain power generation leads to unpredictable fluctuation in both the power demand for traditional power generators and the power flow in the grid system. This may result in serious problems such as voltage instability and cost-ineffective power generation [1]. Energy storage systems (ESSs) offer a promising solution that can significantly contribute to stable and cost-effective supply of electricity energy. Today's ESSs include pumped hydro, various battery technologies (Lead-Acid, Nickel-Cadmium, Sodium-Sulfur, etc.), flywheel, and compressed air energy storage (CAES) [2]. The major problem with pumped hydro and under ground CAES storage is that there are few locations which have the required geological layout that allows these methods to be used. In practice, the locations where pumped hydro/CAES is deployable usually have a long distance from where the electricity is needed. Battery energy storage systems (BESSs), on the other hand, being modular and pad mounted in design, can be put into any traditional electrical sub-stations [3]. Compared with other types of ESSs, BESSs have very fast ramping rates, making them suitable for various system services such as frequency control and real-time market operation.

BESS was used in the early days of direct current electric power [4]. Where AC grid power was not readily available, isolated lighting plants run by wind turbines or internal combustion engines provided lighting and power to small motors. BESS could be used to run the load without starting the engine or when the wind was calm. BESSs that connect to large solid-state converters has been used to stabilize power distribution networks. In recent studies, BESS is considered to be a promising

© The Author(s) 2017
W. Tang, Y.J.A. Zhang, *Optimal Charging Control of Electric Vehicles in Smart Grids*, SpringerBriefs in Electrical and Computer Engineering,
DOI 10.1007/978-3-319-45862-5_1

mechanism for absorbing the variability in local power generation and consumption, thereby reducing the fluctuation in the total power flow [5–10]. The key problem is how to optimally control the charging/discharging of BESSs to maximize its potential benefit for the grid.

The massive integration of plug-in electric vehicles (PEVs) has introduced a new type of BESS, i.e., mobile BESSs [11, 12]. A PEV is a motor vehicle that can be recharged from an external source of electricity, and the electricity stored in the rechargeable battery packs drives or contributes to driving the wheels [13]. The transportation industry represents a major portion of global emission, which is responsible for 24 % of the global carbon dioxide production. The deployment of PEVs on roads is regraded as an efficient way to reduce carbon emission from the transportation industry [14]. As such, world government have pledged billions of dollars to fund the development of PEVs and their components. According to the new analysis from the Centre for Solar Energy and Hydrogen Research, the demand for electric vehicles (EVs) is growing around the world fairly rapidly that brings the total global market more than 740,000 EVs in early 2015 [15].

The fast increasing adoption of EVs brings both challenges and opportunities to the power grid. On one hand, the massive load caused by the integration of EVs into the power grid raises concerns about the potential impacts to the operating cost, voltage stability and the frequency excursion at both generation and transmission sides. On the other hand, EVs can be used as a new type of mobile ESSs that can serve many purposes. With adequate energy stored in the batteries of EVs, the bidirectional charging and discharging control has extensive applications in the microgrids/distribution networks, such as load flattening, peak shaving, frequency fluctuation mitigation and improving the integration of renewable sources. For instance, Fig. 1.1 illustrates the use of EVs for load flattening in a power gird. During off-peak hours, EVs can act as loads to withdraw and store electricity from the main grid. During peak hours, EVs can release the stored energy back to the grid to meet the high demand of other electricity consumers. Overall, the use of EVs flattens the power profile over time and improves the stability of the entire power system. In both cases, uncontrolled EV charging/discharging will lead to inefficient system operation or even severe problems at different network levels. It is therefore critical to develop effective charging/discharging scheduling algorithms for efficient grid operation. In practice, a key design challenge of charging scheduling algorithms lies in the randomness and uncertainty of future events, including the charging profiles of EVs arriving in the future, future load demand in the grid, future renewable energy generation, etc. Therefore, it is necessary to develop online charging/discharging algorithms to cope with different degrees of uncertainty when making real-time decisions. Besides, the large-scale charging of EVs requires low-complexity control mechanisms to reduce the operating delay and the capital cost of equipment investment.

In this book, we study the optimal online charging control of BESSs and PEVs in Microgrids. We first consider the problem of EV charging scheduling, when different amount of future information is available at the time when scheduling decision is made. We then consider the case when bi-directional power flow can

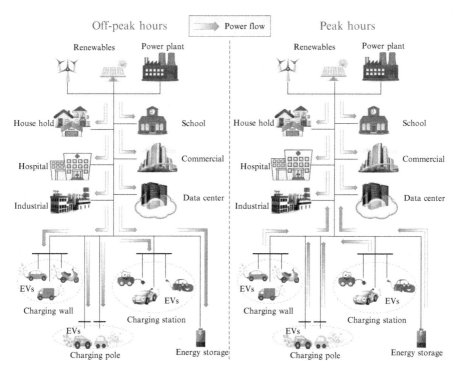

Fig. 1.1   An illustration of the applications of EVs at the time of peak hours and off-peak hours of base load consumptions

be drawn from the BESS to mitigate the impact of renewable energy integration in a microgrid. In all cases, we endeavor to find simple yet optimal or near optimal solutions that are readily deployable in practical systems.

1.2   Backgrounds

1.2.1   Energy Storage Systems

The development of smart power grids has for a long time hinged on concepts of flexibility and demand response [2, 3]. Meanwhile, compared with the traditional power plants, sources like solar or wind tend to output far more uncertain and less controllable power. As such, the balance between supply and demand will depend more on adjusting demand, shifting it over time as required. Eventually, power balance will depend and more often on the availability of buffer/storage capacity available in the network, since there is a limit for individual electrical devices and components to absorb the variability. By providing the energy buffers, power grids could gain additional capacity to absorb the load demand during the peak hours or supply for a longer period of time. Furthermore, the energy capacity will play a crucial role to increase the amount of renewable energy sources in the power grids.

A storage solution is considered a black box that takes in electricity from the grid and releases energy in some form, which naturally influences the balance between demand and supply in the system. This is obvious when energy is released again as electricity and thus adds to the overall supply (e.g. a battery, pumped hydro) [16]. When energy is released/used as heat, it means no electricity needs to be used at that point in time to generate the heat and thus overall demand is suppressed. For example, energy is stored as heat in a heat pump with an integrated thermal buffer, and released when required without the heat pump having to use electricity again [17]. Based on the principle of operation and main components, the energy storage systems can be mainly classified into three types [16–20]:

- Mechanical Storage System,
- Electrochemical storage system,
- Thermal storage system.

Mechanical storage system includes pumped hydro electric system (PHS), flywheel energy storage system (FESS) and compressed air energy storage system (CAES). PHS is a large scale energy storage system, which converts the potential energy of water into electrical energy. FESS is an electromechanical device, which stores the energy in the form of kinetic energy. CAES is capable of providing the large energy storage deliverability of above 100 MW with single unit, which works on the basis of gas-turbine technology.

The typical electrochemical storage system is the battery energy storage technology. The battery technology is one of the oldest storage system, which stores the electrical energy in the form of chemical energy. The battery storage system is comprised of one or more electrochemical cells, where each cell consist of a liquid or solid electrolyte together with a positive and negative electrodes. Based on the electrochemical technologies, the battery storage system can be classified into following types [21]:

- Lead-acid battery storage system
- Lithium-ion battery storage system
- Nickel-cadmium battery storage system
- Sodium-Sulphur battery storage system
- Sodium-nickel Chloride battery storage system
- Vanadium redox flow battery (VRB) storage system
- Zinc bromine battery storage system (Zn Br battery)
- Polysulphide-bromide battery storage system
- Hydrogen fuel cell system

The thermal storage systems are classified into low temperature and high temperature systems [22]. They are further categorised into industrial cooling , building cooling, building heating and industrial heating systems.

1.2.2 Applications of ESSs

The main applications of ESSs in the power system focus on the following aspects [21]:

- Ancillary services: By providing or absorbing energy from the grid, ESSs are able to provide ancillary services including frequency control, voltage control, spinning reserve, standing reserve and black start.
- Peak shaving: The adoption of ESSs can reduce peaks in power demand and therefore effectively reduce the investment cost, because distribution and transmission lines as well as generation capacity are dimensioned according to the peak power demand.
- Load leveling: Load leveling reduces fluctuations in energy demand during one day. During times with low demand, energy is transferred into the ESSs and during times with high demand, the energy stored in the ESSs is fed back into the grid.
- Island grids: In remote areas or on islands, the connection to an integrated power network is in many cases either not economical or technically impossible.
- Other sectors include but not limited to the electromobility, heat storage, residential energy storage for increased self-consumption of distributed electricity generation, industrial energy storage, uninterruptible power supply (UPS).

1.2.3 BESS Models

As the fast development of BESSs, there are different kinds of BESSs for different scenarios, where the installations sizes range from kilowatts to gigawatts, and discharge times range from seconds to hours [2]. Various mathematical models have been developed to predict the operation of BESSs given a set of parameters. There are numerous factors that affect the operation of a battery, including charging/discharging rate, battery age, battery type, temperature, etc. In general, the modelling methods are broadly divided into two types: electrochemical models and equivalent circuit models.

- Electrochemical models of batteries are designed to take into account the chemical, thermodynamic, and physical qualities of the batteries and are typically more precise and complex [23].
- Equivalent circuit models are typically used simulating BESSs in power system applications [24]. The battery voltage, current, charge and temperature vary as functions of one another, which in turn affect the battery output.

1.3   Contributions

This book studies the optimal online charging control of the energy storage system in microgrids and distribution networks. To address some practical concerns of ESS charging control, we pursue the development of the following two design components:

- *Online or stochastic problem formulations* to capture the fact that the information future system data are typically unknown or uncertain at the time when decisions are made;
- *Online algorithm designs* to solve the online or stochastic problems with improved performance in terms of optimality and scalability.

We practise our idea of schemes through the investigation of two types of energy storage systems, i.e., PEVs and BESS. Each of them has some practical challenges, which render the conventional designs infeasible or unscalable. Specifically, the key contributions of this book are summarized into the following three aspects.

- We propose an efficient Online cooRdinated CHARging Decision (ORCHARD) algorithm that aims to minimize the total energy cost without making use of any future information. In contrast to the algorithms proposed in [25–27], ORCHARD allows heterogeneity among PEVs. That is, PEVs can have arbitrary arrival (or plug-in time) and departure times, charging demands and maximum charging rates. We show that ORCHARD is strictly feasible in the sense that it guarantees to fulfill all charging demands before the due time, as long as the charging problem is feasible. More importantly, we rigorously analyze the performance of ORCHARD in terms of competitive ratio, which is a commonly used metric for assessing online algorithms. Our analysis shows that ORCHARD achieves a competitive ratio of 2.39 when the energy cost function is quadratic form of the load demand. This is the best known competitive ratio so far [28]. To further reduce the computational complexity, we propose a low-complexity optimization routine to replace the standard convex optimization algorithms used in ORCHARD. Extensive simulations show that the average performance gap between ORCHARD and the offline optimal solution is as small as 6.5 %. The gap can be reduced to 5 %, if the speeding factor used in the algorithm is carefully chosen according to the charging demand pattern.
- We consider the optimal PEV charging scheduling, assuming that the future charging demand is not known a priori, but its statistical information can be estimated. In particular, we define the cost of PEV charging as a general strictly convex increasing function of the instantaneous load demand. Minimizing such a cost leads to a flattened load demand, which is highly desirable for many reasons [25, 29–32]. The online PEV charging scheduling problem is formulated as a finite-horizon dynamic programming problem with continuous state space and action space. To avoid the prohibitively high complexity of solving such a dynamic programming problem, we rigorously prove that a Model Predictive Control (MPC) approach yields a near-optimal solution that has a bounded

performance gap from the optimal solution regardless of the distribution of exogenous random variables. Specially, the performance gap is evaluated by the Value of the Stochastic Solution (VSS), which represents the gap between the solution of the approximate approach and that of dynamic programming problem [33–35]. Instead of adopting the generic convex optimization algorithms in MPC approach, we propose an algorithm with computational complexity $O(T^3)$ by exploring the load flattening feature of the solution, where T is the total number of time stages. Furthermore, we show that the proposed online algorithm can be made scalable when the random process describing the arrival of charging demands is first-order periodic. That is, the complexity of obtaining the charging schedule at each time stage is reduced to $O(1)$ and is independent of T. Extensive simulations show that the proposed algorithm performs very closely to the optimal solution. The performance gap is smaller than 0.4 % in most cases.

- We address the problem in a microgrid system with renewable energy sources and a BESS. The microgrid exchanges power with an interconnected main grid when its power demand and supply cannot balance internally. To mitigate the negative impact of renewable energy integration, we aim to minimize a cost function, which is a general convex increasing function of the instantaneous power exchanged with the main grid. On one hand, the increasing convexity of objective function reflects the fact that each unit of additional power demand is more expensive to obtain and make available to the consuming entity. On the other hand, it flattens the power exchanged with the main grid over time as much as possible. Through rigorous analysis, we show that the optimal BESS operation policy exhibits a threshold structure, which allows the design of simple control algorithms. When the discount factor satisfies certain conditions, the optimal policy degenerates to one that takes a *short-sighted* behavior, i.e., to discharge the battery as fast as possible regardless of the system state. This short-sighted policy drains the battery as fast as possible without recharging it, making BESS almost useless. As such, our analysis here provides a guideline to set the discount factor such that the short-sighted policy is avoided. Moreover, we discuss the effect of the battery energy capacity on the total cost under the optimal charging policy. We show that the optimal cost is a decreasing convex function of the battery capacity, implying that there exists an optimal battery sizing that strikes a balance between the total cost and the capital investment. Our simulation results show that the battery can significantly reduce the large fluctuation in the power demand caused by the integration of renewable energy sources. The numerical results also verify our analysis, including the threshold structure of the optimal policy, the condition when the short-sighted policy occurs, and the effect of the battery energy capacity.

1.4 Organization

This book is organized as follows. Chapter 2 investigates the PEV charging scheduling problem without any assumptions or predictions of the future information. An Online cooRdinated CHARging Decision (ORCHARD) algorithm is proposed and its worst case performance is quantified in terms of competitive ratio. In Chap. 3, we consider the PEV charging scheduling problem with the estimation of statistical information of random data. A Model Predictive Control (MPC) based algorithm is proposed and analyzed in terms of both optimality and scalability. Chapter 4 is concerning the optimal control of a battery energy storage system (BESS) in a microgrid with renewable energy sources. We show that the optimal charging policy has a threshold structure. Besides, the optimal battery size is analyzed under the optimal charging policy. Finally, we conclude this book in Chap. 5 by summarizing the main results of this book and discuss several potential directions in future work.

References

1. J. Twidell and T. Weir, *Renewable energy resources*, Routledge, 2015.
2. D. Rastler, "Electricity energy storage technology options, a white paper primer on applications, costs and benefits," EPRI technical update, 2010.
3. B. Dunn, H. Kamath, and JM. Tarascon, "Electrical energy storage for the grid: a battery of choices," *Science*, 334.6058, pp. 928–935, 2011.
4. Power Sonic, *Sealed Lead-Acid Batteries Technical Manual*, 2014.
5. J. Qin, R. Sevlian, D. Varodayan, and R. Rajagopal, "Optimal electric energy storage operation," *in Proc. of IEEE PES General Meeting*, pp. 1–6, Jul. 2012.
6. P. van de Ven, N. Hegde, L. Massoulie, and T. Salonidis, "Optimal control of end-user energy storage," *IEEE Trans. on Smart Grid*, vol. 4, no. 2, pp. 789–797, 2013.
7. Y. Xu and L. Tong, "On the operation and value of storage in consumer demand response," *53rd IEEE Conference on Decision and Control*, pp. 205–210, Dec. 2014.
8. Y. Ru, J. Kleissl, and S. Martinez, "Storage size determination for grid-connected photovoltaic systems," *IEEE Trans. on Sustainable Energy*, vol. 4, no. 1, pp. 68–81, 2013.
9. I. Koutsopoulos, V. Hatzi, and L. Tassiulas, "Optimal energy storage control policies for the smart power grid," *Proc. of IEEE International Conference on Smart Grid Communications (SmartGridComm)*, pp. 475–480, Oct. 2011.
10. L. Huang, J. Walrand, and K. Ramchandran, "Optimal demand response with energy storage management," *Proc. of IEEE International Conference on Smart Grid Communications (SmartGridComm)*, pp. 61–66, Nov. 2012.
11. S. S. Hosseini, A. Badri, and M. Parvania, "A survey on mobile energy storage systems (MESS): Applications, challenges and solutions," *Renewable Sustainable Energy Rev. 40*, pp. 161–170, 2014.
12. S. S. Hosseini, A. Badri, and M. Parvania, "The plug-in electric vehicles for power system applications: the vehicle to grid (V2G) concept," *Proc. IEEE Int. Energy Conf. Exhib.*, pp. 1101–1106, Sept. 2012.
13. N. R. E. L. (NREL), "Electric Vehicle Grid Integration," 2015.
14. A. Y. Saber and G. K. Venayagamoorthy, "Plug-in vehicles and renewable energy sources for cost and emission reductions," *IEEE Trans. on Industrial Electronics*, vol. 58, no. 4, pp. 1229–1238, 2011.

15. A. Regno and A. Vartmann, "More than 740,000 cars worldwide powered by electricity," The Centre for Solar Energy and Hydrogen Research, Stuttgart, 2015.
16. J. Cao and A. Emadi, "A new battery/ultracapacitor hybrid energy storage system for electric, hybrid, and plug-in hybrid electric vehicles," *IEEE Trans. Power Electron.*, vol. 27, no. 1, pp. 122–132, 2012.
17. F. DAaz-GonzAilez, A. Sumper, O. Gomis-Bellmunt, and V. Robles, "A review of energy storage technologies for wind power applications," *Elsevier Renewable Sustain. Energy Rev.*, vol. 16, no. 4, pp. 2154–2171, 2012.
18. A. Khaligh and Z. Li, "Battery, ultracapacitor, fuel-cell, and hybrid energy storage systems for electric, hybrid electric, fuel cell, and plug-in hybrid electric vehicles: State-of-Art," *IEEE Trans. Veh. Technol.*, vol. 59, no. 6, pp. 2806–2814, 2010.
19. H. Ibrahim, A. Ilinca, and J. Perron, "Energy storage systems-characteristics and comparisons," *Renewable Sustainable Energy Rev.*, vol. 12, no. 5, pp. 1221–1250, 2008.
20. J. M. Carrasco, L. G. Franquelo, J. T. Bialasiewicz, E. Galvan, R. C. P. Guisado, M. A. M. Prats, J. I. Leon, and N. Moreno-Alfonso, "Power-electronic systems for the grid integration of renewable energy sources: A survey," *IEEE Trans. Power Electron.*, vol. 53, no. 4, pp. 1002–1016, 2006.
21. G. Fuchs, B. Lunz, M. Leuthold, and D. U. Sauer, "Technology Overview on Electricity Storage," *RWTH Aachen*, 2012.
22. H. L. Ferreira, R. Garde, G. Fulli, W. Kling, and J.P. Lopes, "Characterisationof electrical energy storage technologies," *Energy*, vol. 53, pp. 288–298, 2013.
23. D. Cadar, D. Petreus, I. Ciocan and P. Dobra, "An Improvement on Empirical Modelling of the Batteries," *32nd International Spring Seminar on Electronics Technology (ISSE 2009)*, pp. 1–6, 2009.
24. K. Yoon-Ho and H. Hoi-Doo, "Design of interface circuits with electrical battery models," *IEEE Trans. Ind. Electron.*, vol. 44, no. 1, pp. 81–86, 1997.
25. Z. Ma, D. Callaway, and I. Hiskens, "Decentralized Charging Control of Large Populations of Plug-in Electric Vehicles", *IEEE Trans. on Control Systems Technology*, vol.21, no.1, pp. 67–78, 2013.
26. K. Clement-Nyns , E. Haesen, and J. Driesen, "The impact of charging plug-in hybrid electric vehicles on a residential distribution grid," *IEEE Trans. Power Syst.*, vol. 25, no. 1, pp. 371–380, 2010.
27. M. A. S. Masoum, P. S. Moses, and S. Hajforoosh, "Distribution transformer stress in smart grid with coordinated charging of plug-in electric vehicles," *IEEE Power Energy Syst. Innovative Smart Grid Tech. Conf.*, pp. 1–8, 2012.
28. N. Bansal, H. L. Chan, K. Pruhs, and D. Katz, "Improved bounds for speed scaling in devices obeying the cube-root rule," *Proc. 36th Int. Colloqium on Automata, Languages and Programming: Part I*, pp. 144–155, Jul. 2009.
29. E. Sortomme, M. M. Hindi, S. D. J. MacPherson, and S. S. Venkata, "Coordinated charging of plug-in hybrid electric vehicles to minimize distribution system losses," *IEEE Trans. Smart Grid*, vol.2, no.1, pp. 198–205, 2011.
30. W. Tang and Y. J. Zhang, "A model predictive control approach for low-complexity electric vehicle charging scheduling: optimality and scalability," accepted by *IEEE Trans. Power Syst.*, Jun. 2016.
31. Y. He, B. Venkatesh, and L. Guan, "Optimal scheduling for charging and discharging of electric vehicles," *IEEE Trans. on Smart Grid*, vol.3, no.3, pp. 1095–1105, 2012.
32. L. Gan, U. Topcu, and S. H. Low, "Optimal decentralized protocol for electric vehicle charging," *IEEE Trans. on Power System*, vol.28, iss. 2, pp. 940–951, 2012.
33. J. R. Birge and F. Louveaux, *Introduction to Stochastic Programming*, New York: Springer, 1997.

34. B. Defourny, D. Ernst, L. Wehenkel, L. E. Sucar, E. F. Morales, and J. Hoey, "Multistage stochastic programming: A scenariotree based approach to planning under uncertainty," *in DecisionTheory Models for Applications in Artificial Intelligence: Concepts and Solutions*, pp. 51, 2011.
35. F. Maggioni and S. Wallace, "Analyzing the quality of the expected value solution in stochastic programming," *Annals of Operations Research*, pp. 37–54, 2012.

Chapter 2
ORCHARD Algorithm for PEV Charging

In this chapter, we consider the PEV charging problem in a community, where the power consumption consists of the load of a PEV charging station and the other inelastic base load, as shown in Fig. 2.1. By controlling the charging rates of PEVs, we aim to minimize total cost on electricity bill paid by the charging station. PEVs arrive at the charging station at random instants with random charging demands that must be fulfilled before their departure time.

The optimal PEV charging scheduling problem has been widely studied in the literature. Many of the existing PEV charging algorithms are "offline" in the sense that they rely on the non-causal information of future PEV charging profiles when deciding the charging schedules. That is, the arrival time and charging demand of a PEV are assumed to be known to the charging station prior to the arrival of the PEV. For instance, [1] requires all PEVs to negotiate with the charging station about their charging schedules one day ahead. However, this assumption does not hold in practice. A PEV's charging profile is revealed only after it arrives at the charging station or connects to the charging pole. Considering the most conservative case where neither the future PEV arrival instants, charging demands, nor their distributions are known *a priori*, we are interested in developing an *online* charging algorithm that schedules PEV charging based only on the information of the PEVs that have already arrived at the charging station.

2.1 Problem Formulation

In this section, we first introduce the offline PEV charging problem by assuming the knowledge of future information. We then formulate the online PEV charging problem without future knowledge. The optimal offline PEV charging scheme

© The Author(s) 2017

W. Tang, Y.J.A. Zhang, *Optimal Charging Control of Electric Vehicles in Smart Grids*, SpringerBriefs in Electrical and Computer Engineering, DOI 10.1007/978-3-319-45862-5_2

Fig. 2.1 Illustration of PEV
charging scenario

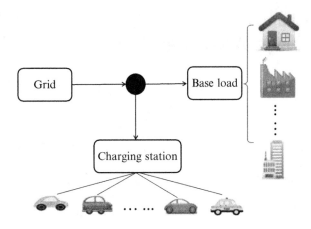

will be used as a benchmark to evaluate the performance of the proposed online
algorithm.

2.1.1 Optimal Offline PEV Charging Problem

Suppose that N PEVs arrive during a time period T, indexed from 1 to N according
to their arrival order. Notice that for a given time period T, N itself is a random
variable due to the random arrival of PEVs. Let $D_i, t_i^{(s)}, t_i^{(e)}$ denote the charging
demand, arrival time, and departure time of PEV i, respectively, which will be
known by the charging station once the PEV arrives. In order to capture the key
characteristic of the online charging problem, we assume that a PEV will not depart
unless its charging demand is fulfilled. Nonetheless, later we will show that our
online algorithm is not affected by early charging terminations.

Due to the battery constraint, PEV i can only be charged at a rate $x_{it} \in [0, U_i]$,
where U_i is the maximum charging rate. For the formulation to be meaningful, we
assume that all the charging demands are feasible. That is,

$$D_i \leq \min\{U_i(t_i^{(e)} - t_i^{(s)}), \zeta_i\} \tag{2.1}$$

holds for all i, where ζ_i is the battery capacity of PEV i. For simplicity, we omit the
upper bound of the total charging rate that can be provided by the charging station.
Let $\mathscr{I}_t$ be the set of PEVs parking in the station at time t. The charging station has
the control of the charging rate x_{it} for each PEV i. We define s_t as the total charging
rate at time t, i.e.,

$$s_t = \sum_{i \in \mathscr{I}_t} x_{it}, \tag{2.2}$$

which is also called charging load at time t. The total load consists of the charging load and the inelastic base load. The base load, denoted by l_t, represents the load of other electricity consumptions at time t except for PEV charging. Here, we assume that the base load does not change continuously with time. Rather, it remains constant for a duration of time (usually in the unit of seconds or minutes) and varies to another value afterwards (see Fig. 2.2 for the illustration). Then, the total load at time t, denoted by y_t, is given by

$$y_t = s_t + l_t = \sum_{i \in \mathscr{I}_t} x_{it} + l_t. \tag{2.3}$$

In this chapter, we assume that the community pays a wholesale electricity price that is time-varying and determined by the total power consumption rate in the system. This often corresponds to a generator supporting a small geographic area with only the temporal variation but no spatial variation of the price [1, 2]. The electricity price is modeled as a linear function of the instant load [1, 3], which is given as follows:

$$a + 2bz_t, \tag{2.4}$$

where a and b are non-negative real numbers, z_t is the instant load. Similar to [3], the electricity cost paid by the charging station at time t is given by

$$\int_{l_t}^{y_t} (a + 2bz_t) dz_t = (a(\sum_{i \in \mathscr{I}_t} x_{it} + l_t) + b(\sum_{i \in \mathscr{I}_t} x_{it} + l_t)^2) - (al_t + bl_t^2), \tag{2.5}$$

which indicates that the charging station should be responsible for the increased electricity cost caused by the PEV charging. The total cost paid by the charging station for the electricity bill within $[0, T]$ is denoted by Ψ and computed by

$$\Psi = \int_0^T \left(a(\sum_{i \in \mathscr{I}_t} x_{it} + l_t) + b(\sum_{i \in \mathscr{I}_t} x_{it} + l_t)^2 - (al_t + bl_t^2) \right) dt. \tag{2.6}$$

The optimal charging scheduling problem that minimizes the total energy cost is then formulated as (2.7).

$$\min_{x_{it}} \quad \int_0^T \left(a(\sum_{i \in \mathscr{I}_t} x_{it} + l_t) + b(\sum_{i \in \mathscr{I}_t} x_{it} + l_t)^2 - (al_t + bl_t^2) \right) dt \tag{2.7a}$$

$$\text{s. t.} \quad \int_{t_i^{(s)}}^{t_i^{(e)}} x_{it} dt = D_i, i = 1, 2, \ldots, N, \tag{2.7b}$$

$$0 \le x_{it} \le U_i, i = 1, 2, \ldots, N, t \in \left[t_i^{(s)}, t_i^{(e)} \right]. \tag{2.7c}$$

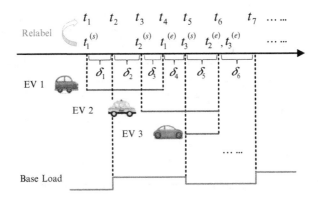

Fig. 2.2 Illustration of one offline case. The time instants are defined as the arrivals and departures of PEV $1, 2, 3, \cdots$, and the change times of base load. Then, relabel the time instants as $t_1, t_2, \cdots$ in a sequential order. In this case, both PEV2 and PEV3 leave at time t_6, and both PEV 3 arrives and base load changes at time t_5

As shown in [4], (2.7) also captures the intent of flattening the total load over time since the cost function is a convex function of total load. It is obvious that Problem (2.7) is a convex optimization problem. In the ideal case where the base load l_t and all PEVs' charging profiles, including $t_i^{(s)}$, $t_i^{(e)}$, U_i and D_i are known to the charging station noncausally at time 0, one can obtain the optimal x_{it} for all i and t by solving (2.7) before the start of system time. We refer to the optimal solution obtained with noncausal information as the offline optimal solution. In practice, however, a PEV's charging profile is revealed only after it arrives at the station. Meanwhile, the base load is also a time-varying random process that cannot be precisely predicted beforehand. In Sect. 2.3, we will investigate an online PEV charging problem that determines the charging rate at each time t based only on the current and past information.

2.1.2 Model Transformation

A close look at (2.7) suggests that there are infinite number of variables x_{it}, because the time t is continuous. In this subsection, we show that the problem (2.7) can be equivalently transformed to a discrete model that is easier to solve and more practical to implement, i.e., the optimal charging rate changes only once in a while.

The equivalence is established through transforming the original continuous problem (2.7) to an event-driven discrete time problem. Throughout this chapter, an *event* is defined by an PEV arrival, departure, or a change in the base load. Likewise, a time interval is defined as the time period between two adjacent events. As illustrated in Fig. 2.2, we relabel the time instants when the events occur as $t_1, t_2, \cdots$ in a sequential order. By doing so, neither the base load nor the set of

PEVs parked in the station changes in the middle of a time interval. Here, we does not exclude the possibility that more than one event occurs at the same time. For instance, in Fig. 2.2, both PEV2 and PEV3 leave at time t_6, and both PEV 3 arrives and base load changes at time t_5. Let $\mathcal{K}$ denote the set of indices of the time intervals, and $\delta_k (k \in \mathcal{K})$ denote the length of kth interval. Without loss of generality, we denote the base load during the kth interval $[t_k, t_{k+1})$ by l_k, since it does not change within a time interval. We show in Lemma 2.1 that there exists an optimal solution where the charging rates remain constant during each time interval.

Lemma 2.1. *Let x_{it}^* denote an optimal solution to (2.7) and $s_t^* = \sum_{i \in \mathcal{I}_t} x_{it}^*$. Then, the optimal total charging rate s_t^* remains constant in each time interval. Moreover, there exists an optimal solution where x_{it}^* remains constant during each time interval.*

The lemma can be easily proved by Jensen's inequality. The detailed proof please refer to Appendix "Proof of Lemma 2.1".

Due to Lemma 2.1, we can safely assume that x_{it} do not change during a time interval. Denote by x_{ik} the charging rate of PEV i during the kth interval. Likewise, denote $\mathcal{J}(i)$ as the set of indices of the time intervals during which PEV i parks in the station, $\mathcal{I}(k)$ as the set of PEVs that park in the kth interval. Based on Lemma 2.1, we can equivalently transform problem (2.7) to the following form that has finitely many variables:

$$
\min_{x_{ik}} \quad \sum_{k \in \mathcal{K}} \left(a\Big(\sum_{i \in \mathcal{I}(k)} x_{ik} + l_k \Big) + b\Big(\sum_{i \in \mathcal{I}(k)} x_{ik} + l_k \Big)^2 \right.
$$
$$
\left. - (al_k + bl_k^2) \right) \delta_k \tag{2.8a}
$$

$$
\text{s.t.} \quad \sum_{k \in \mathcal{J}(i)} x_{ik}\delta_k = D_i, i = 1, 2, \dots, N, \tag{2.8b}
$$

$$
0 \le x_{ik} \le U_i, i = 1, 2, \dots, N, k \in \mathcal{J}(i). \tag{2.8c}
$$

It is worth pointing out that the discrete time model in (2.8) is different from the traditional time-slotted models. The lengths of time slots are fixed in traditional time-slotted models, whereas the variables in (2.8) are defined by the random events. By doing so, the model in (2.8) successfully captures the dynamics in the system, which is not achievable by the traditional time-slotted models unless the time slots are set infinitesimally small.

2.1.3 Online PEV Charging Problem and Performance Metric

The online PEV charging problem assumes that, at any time instant t, the scheduler only knows the information that is available so far, including the charging profiles of the PEVs that have arrived upon or before t, as well as the past and current base load. Based on the causal information, the scheduler makes an online decision of the charging rates x_{it} when an event occurs, and the charging rates remain unchanged until the occurrence of the next event. Notice that for practicality, a past decision that has already been implemented cannot be changed in the future. Thus, without knowing the future information, an online algorithm is forced to make decisions that may later turn out to be suboptimal. That is, we have $\Psi_{ON} \geq \Psi^*$, where Ψ_{ON} denotes the total cost induced by an online algorithm and Ψ^* denotes the optimal cost obtained by the offline optimization.

A standard metric to evaluate the performance of an online algorithm is the competitive ratio, which compares the relative performance of an online and the offline algorithm under the same sequence of inputs (e.g., the PEV charging profiles in our problem)[5]. In particular, the competitive ratio of an online algorithm is the maximum ratio between its performance and that of the offline optimal algorithm over all possible input sequences. The formal definition is given in the following Definition 2.1 [5].

Definition 2.1. An online algorithm is c−competitive if there exists a constant θ such that

$$\Psi_{ON} \leq c \cdot \Psi^* + \theta \tag{2.9}$$

holds for any input.

By definition, the competitive ratio is always greater than or equal to 1. Notice that the competitive ratio measures the performance ratio in the worst case. Very often, the average performance ratio is much smaller than c. This will be shown in the simulation section, where the proposed ORCHARD algorithm achieves an average performance ratio less than 1.06, although the competitive ratio is 2.39 when the cost function is a quadratic function of load demand.

2.2 Related Work

There have been some recent studies on online PEV charging [3, 4, 6–11]. Gerding et al. [6] proposes an online auction protocol that vehicle owners use agents to bid for the charging opportunities. Therein, it assumes that all the PEVs have the same fixed charging rate. In practice, however, the charging rate could vary among different types of PEVs. Masoum et al. [7] studies the coordinated charging of PEVs in residential distribution systems to reduce the power loss, by assuming that all the PEVs have the same charging period. In practice, the PEVs are very likely to be at the charging station during different time periods. In contrast to the algorithms proposed in [6–8], the proposed PEV charging algorithms in this chapter allow heterogeneity among PEVs. That is, PEVs can have arbitrary arrival (or plug-in time) and departure times, charging demands and maximum charging rates. He et al. [3] considers the scheduling of PEV charging and discharging in a small geographic area and proposes an online charging algorithm based on an assumption that no future PEV will arrive when a charging schedule is made. The resulting charging schedule is suboptimal as it underestimates the actual charging load. More importantly, most of the existing work, including [3, 4, 6–8], do not provide theoretical analysis of their online algorithms. The few works that analyze the performance, e.g., [9], do not guarantee the satisfaction of PEVs' charging demands before their departures. Compared with [3, 4, 6–9], we have rigorously analyzed the performance of the proposed PEV charging algorithms and show that the proposed PEV charging algorithms are strictly feasible in the sense that they guarantees to fulfill all charging demands before the due time. In addition, the cost functions adopted in [10, 11] depend on the individual PEV's charging demand, whereas in this chapter we consider the price based on the aggregate load demand, including the charging demand of all PEV users as well as the base load.

The charging scheduling for PEV is similar to, but not the same as, the speed scaling problem, which is a power management technique that involves dynamically changing the speed of a processor [12–16]. Specifically, the processor must schedule in real-time a number of tasks and allocate a processing rate to each of them, given that all tasks can be completed before their predetermined deadlines. The processor tries to minimize the total energy cost, where the energy cost at each time t is a positive power function of the total processing rate $s(t)$ at that time, i.e. $s^\alpha(t)$ and $\alpha > 1$. The key difference from a PEV charging problem is that speed scaling studied in [12–14] does not place a constraint on the maximum processing rate of each individual job as the PEV charging problem, i.e., each PEV has a maximum charging rate. Another difference is that the cost function of PEV charging problem is a general polynomial instead of a positive power function. In other words, PEV charging schedule is a more general problem than the speed scaling problem, thus its competitive ratio is no less than 2.39, i.e., the best known ratio for speed scaling problem when the cost function is a quadratic form [14].

The first offline optimal algorithm to solve the speed scaling problem was proposed by Yao, Demers and Shenker (YDS) [12]. Later, [12] proposed two online algorithms, i.e. Average Rate (AVR) and Optimal Available (OA). Conceptually, AVR processes a task at a rate equals to its average work load within its specified starting time and deadline. The algorithm is proved to be $2^{\alpha-1}\alpha^\alpha$-competitive in [12]. OA uses YDS to calculate the current optimal processing rate by assuming no more tasks will be released in the future, and its competitive ratio was proved to be α^α [13]. Apparently, the OA solution is suboptimal, as it underestimates the future workload. To address the problem, [14] proposed a qOA algorithm that scales up the processing rate of OA by a factor $q > 1$. It also showed that qOA works better than OA and AVG in terms of competitive ratio. There are many follow-up works on extended topics, such as managing both temperature and power [13], minimizing the total flow plus energy [15, 16] , etc. Overall, the existing online algorithms for speed scaling cannot be directly applied to solve our problem, mainly because they do not consider the limits on the maximum processing speeds of tasks.

In contrast to the previous work, we propose an Online cooRdinated CHARging Decision (ORCHARD) algorithm, which minimizes the energy cost by mimicking the offline optimal charging decision. Note that ORCHARD relies on no assumptions or predictions of the future information. Through rigorous proof, we show that ORCHARD is strictly feasible in the sense that it guarantees to fulfill all charging demands before due time. Meanwhile, it achieves the best known competitive ratio of 2.39, when the cost function is a quadratic function of the load demand. By exploiting the problem structure, we propose a novel reduced-complexity algorithm to replace the standard convex optimization techniques used in ORCHARD. Through extensive simulations, we show that the average performance gap between ORCHARD and the offline optimal solution, which utilizes the complete future information, is as small as 6.5 %. By setting a proper speeding factor, the average performance gap can be further reduced to 5 %.

2.3   Online Algorithm

In this section, we present an efficient online algorithm ORCHARD. We show that ORCHARD achieves a competitive ratio that is the best known so far. Moreover, the algorithm is strictly feasible in the sense that it always ensures to satisfy all PEV charging demands.

The proposed ORCHARD algorithm could be easily implemented in a practical charging station. On one hand, it has low computational complexity. On the other hand, it only relies on the causal information of the vehicles and the base load rather than the schedule of the vehicles and the base load in the future. It is robust under any PEV traffic distribution and base load pattern, because it involves no predictions about the future information of PEVs and the base load.

2.3.1   Online Optimal Available (OA) Algorithm

In this subsection, we describe a simple online scheme called Optimal Available (OA) algorithm, which, although suboptimal, will be helpful later in understanding our proposed ORCHARD algorithm.

The OA algorithm works as follows. At a time instant t_j when an event occurs, the scheduler calculates the optimal charging schedule assuming that no more PEVs will arrive and base load is unchanged in the future. More specifically, the scheduler solves the following problem (2.11) at time instant t_j, where $\bar{\mathscr{I}}(t, t_j)$ denotes the set of PEVs who have arrived by time t_j and will be in the station at time $t \in (t_j, \bar{T}(t_j)]$, where

$$\bar{T}(t_j) = \max\{t_i^{(e)} : i \in \mathscr{I}_{t_j}\} \tag{2.10}$$

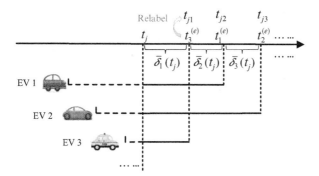

Fig. 2.3 Illustration of one online case. The time intervals are defined by change times of base load as well as the departures of PEVs that are parking in the station at current time t_j. Then, relabel the time instants as $t_{j1}, t_{j2}, \cdots$ in a sequential order

denotes the latest departure time of all PEVs that have already arrived by t_j (recall that $\mathscr{I}_{t_j}$ is the set of PEVs parking in the station at time t_j), $\bar{D}_i(t_j)$ denotes the residual demand to be satisfied for PEV i at time t_j, i.e., the unfinished charging demand of PEV i observed at time t_j.

$$\min_{x_{it}} \quad \int_{t_j}^{\bar{T}(t_j)} \left(a\left(\sum_{i \in \mathscr{I}(t,t_j)} x_{it} + l_{t_j} \right) + b\left(\sum_{i \in \mathscr{I}(t,t_j)} x_{it} + l_{t_j} \right)^2 \right.$$

$$\left. - (al_{t_j} + bl_{t_j}^2) \right) dt \tag{2.11a}$$

$$\text{s. t.} \quad \int_{t \in [t_j, t_i^{(e)}]} x_{it} dt = \bar{D}_i(t_j), i \in \mathscr{I}_{t_j}, \tag{2.11b}$$

$$0 \le x_{it} \le U_i, i \in \mathscr{I}_{t_j}, t \in \left[t_j, t_i^{(e)} \right]. \tag{2.11c}$$

Having obtained the solution to (2.11), the scheduler charges the PEVs according to the solution until a new PEV arrives or the base load changes. Then, Problem (2.11) is re-solved with the updated set of charging profiles and base load level. Similar to the discussion in Section II-B, the time axis (from t_j to $\bar{T}(t_j)]$) in Problem (2.11) can be divided into intervals, which are defined only by the departures of the existing PEVs, since the events occurred after time t_j, i.e., the arrivals/departures of new PEVs or the changes of base load after time t_j are not known by the scheduler. By keeping a charging rate in each interval constant, Problem (2.11) can be equivalently transformed to one with finitely many variables. An example in Fig. 2.3 illustrates the intervals defined by change times of base load and the departures of PEVs that are parking in the station at time t_j. Denote $\mathscr{K}(t_j)$ as the set of indices of the intervals seen at time t_j, $\bar{\delta}_k(t_j)$, where $k \in \mathscr{K}(t_j)$ as the length of the kth interval, $\bar{\mathscr{I}}(k, t_j)$ as the set of PEVs who have arrived by time t_j

and will be in the station at interval $k, k \in \mathscr{K}(t_j)$, and $\bar{\mathscr{J}}(i, t_j)$ as the set of indices of time intervals that PEV i will park in the station. It directly follows from Lemma 2.1 that there exists an optimal solution to (2.11) where the optimal charging rates are constants during each interval. Denote x_{ik} by the charging rate of PEV i in interval $k, k \in \mathscr{K}(t_j)$. Then, (2.11) is equivalent to the following discrete time optimization problem

$$\min_{x_{ik}} \quad \sum_{k \in \mathscr{K}(t_j)} \left(a\left(\sum_{i \in \bar{\mathscr{J}}(k,t_j)} x_{ik} + l_{t_j} \right) + b\left(\sum_{i \in \bar{\mathscr{J}}(k,t_j)} x_{ik} + l_{t_j} \right)^2 \right.$$

$$\left. - (al_{t_j} + bl_{t_j}^2) \right) \bar{\delta}_k(t_j) \tag{2.12a}$$

$$\text{s.t.} \quad \sum_{k \in \bar{\mathscr{J}}(i,t_j)} x_{ik} \bar{\delta}_k(t_j) = \bar{D}_i(t_j), i \in \mathscr{I}_{t_j}, \tag{2.12b}$$

$$0 \le x_{ik} \le U_i, i \in \mathscr{I}_{t_j}, k \in \bar{\mathscr{J}}(i, t_j). \tag{2.12c}$$

In the next section, we will introduce our proposed ORCHARD algorithm. Note that ORCHARD also solves Problem (2.12), but only uses x_{i1}, $i \in \mathscr{I}(1, t_j)$, i.e., the charging solutions in the first (i.e., current) interval. As we will introduce later, (2.12) needs to be resolved again with the updated l_{t_j}, $\bar{\delta}_k(t_j)$, $\bar{\mathscr{J}}(k, t_j)$, $\bar{\mathscr{J}}(i, t_j)$, $\bar{D}_i(t_j)$ once a new PEV arrives, finishes charging, or the base load changes.

2.3.2 The ORCHARD Algorithm

The charging rate scheduled by OA tends to be smaller than the optimal offline solution due to the neglect of future demands. In ORCHARD, we speed up the charging schedule obtained from (2.12) by a speed-up factor q ($q \ge 1$). Roughly speaking, the total charging rate by ORCHARD is q times that of OA. The value of q determines the performance of ORCHARD, including both the competitive ratio and the average performance. We will discuss how to set a proper q to obtain the minimum competitive ratio in Sect. 2.3.3 and to obtain the best average performance in Sect. 2.5.3.

Due to the factor q, the charging rate of ORCHARD is larger than that of OA such that ORCHARD finishes charging PEVs earlier than OA does. Then ORCHARD always finishes charging PEVs before their departure time. Hence, ORCHARD recalculates the charging rate when there is a new PEV arrival, a PEV finishes charging, or the base load changes. We denote by $\bar{x}_{ik}(t_j)$ the charging rate of PEV $i \in \mathscr{I}_{t_j}$ in the kth interval computed by OA at time t_j, $\hat{x}_{it}$ the charging rate of PEV i at time t computed by ORCHARD, and $\hat{s}_t$ the sum of $\hat{x}_{it}$ at time t. When ORCHARD recalculates the charging rate, the right hand side of (2.12b) is updated as follows

$$\bar{D}_i(t_j)$$

$$= \begin{cases} 0, & \text{if PEV } i \text{ finishes charging,} \\ D_i, & \text{if PEV } i \text{ arrives,} \\ \bar{D}_i(t_{j-1}) - \hat{x}_{it_{j-1}}(t_j - t_{j-1}), & \text{otherwise.} \end{cases} \quad (2.13)$$

Here, $\hat{x}_{it_{j-1}}$ denotes the constant charging rate between t_{j-1} and t_j. Moreover, we also need to update $\bar{\mathcal{J}}(i,t_j)$, $\bar{\delta}_k(t_j)$ and $\bar{\mathcal{J}}(k,t_j)$ due to the change of current PEVs at t_j. A pseudo code of ORCHARD is presented in Algorithm 1 and explained as follows:

Step 1: Once the base load changes, a PEV arrives, or a PEV finishes charging, let $j = j + 1$, set t_j as the current starting time and update the current base load l_{t_j} as well as the parameters based on current PEVs (line 3).

Step 2: Solve (2.12) with the updated parameters. Denote by $\bar{x}_{i1}(t_j)$ the optimal charging solution of the first (i.e., current) time interval. (line 4).

Step 3: Determine the total charging rate, which is the minimum of q times of the total charging rate computed by OA, i.e., $\sum_{i \in \mathcal{I}_{t_j}} \bar{x}_{i1}(t_j)$, and the sum of maximum charging rates of current PEVs, i.e., $\sum_{i \in \mathcal{I}_{t_j}} U_i$ (line 5).

Step 4: Determine the charging solution at time $[t_j, t_{j+1})$ by setting the charging rate of PEV i as in line 6 in Algorithm 1, where t_{j+1} is the next time that the base load changes, a PEV arrives, or a PEV finishes charging.

By doing so, we ensure that: (1) for each PEV, the charging rate does not exceed its maximum charging rate, i.e., $x_{it} \leq u_i, i \in \mathcal{I}_{t_j}$; (2) the sum of the charging rates equals the total charging rate given by Step 3, i.e., $\sum_{i \in \mathcal{I}_{t_j}} \hat{x}_{it} = \hat{s}_t$; (3) for each PEV, the charging rate is no smaller than the solution given by OA in Step 2, i.e., $\hat{x}_{it} \geq \bar{x}_{i1}(t_j), \forall i \in \mathcal{I}_{t_j}$ (line 6).

Algorithm 1: ORCHARD

input : U_i, $t_i^{(e)}$, D_i of all parking PEVs, the base load l_t
output: $\hat{x}_{it}$
1 initialization $j = 0$;
2 **while** *the base load changes, a PEV arrives, or a PEV finishes charging* **do**
3   Let $j = j + 1$, record current time t_j. Update l_{t_j}, $\bar{\delta}_k(t_j)$, $\bar{\mathcal{J}}(k,t_j)$, $k \in \mathcal{K}(t_j)$, $\bar{\mathcal{J}}(i,t_j)$, $\bar{D}_i(t_j)$, $i \in \mathcal{I}_{t_j}$.
4   Solve problem (2.12) for the optimal solution $\bar{x}_{i1}(t_j) \forall i \in \bar{\mathcal{I}}_{t_j}$.
5   Set $\hat{s}_t = \min\{q \cdot \sum_{i \in \mathcal{I}_{t_j}} \bar{x}_{i1}(t_j), \sum_{i \in \mathcal{I}_{t_j}} U_i\}$.
6   Set the charging rate of PEV i at the time $t \in [t_j, t_{j+1})$ as
    $\hat{x}_{it} = \min\{\bar{x}_{i1}(t_j) + \frac{U_i - \bar{x}_{i1}(t_j)}{\sum_{i \in \mathcal{I}_{t_j}} (U_i - \bar{x}_{i1}(t_j))} \cdot \frac{q-1}{q} \hat{s}_t, U_i\}$.

7 **end**

Since OA always guarantees a feasible solution, we can intuitively infer that
ORCHARD also guarantees producing a feasible solution, simply because its
charging rate is always no smaller than that of the OA. The feasibility of ORCHARD
is proved in Lemma 2.2 below.

Lemma 2.2. *ORCHARD always outputs a feasible solution to problem (2.12).*

Proof. Please see the detailed proof in Appendix "Proof of Lemma 2.2".   ∎

2.3.3   Derivation of Competitive Ratio

In this subsection, we show that ORCHARD is 2.39-competitive when the cost
function is a quadratic function of the load demand. Here, we consider an amortized
local competitiveness analysis and a potential function $\Phi(t)$ as a function of time.
In the following, we will construct a $\Phi(t)$ to prove the inequality (2.9) for a specific
competitive ratio c. In particular, Φ is chosen to satisfy

$$\Phi(0) = \Phi(T) = 0. \tag{2.14}$$

We always denote the current time as τ_0. Let $l, \hat{s}$ and s^* be the current base load,
total charging rate of ORCHARD and the optimal offline algorithm respectively. In
order to establish that ORCHARD is c-competitive, it is sufficient to show that the
following key equation

$$(a(\hat{s}+l)+b(\hat{s}+l)^2-(al+bl^2))+\frac{\mathrm{d}\Phi}{\mathrm{d}\tau_0}$$
$$\leq c \cdot (a(s^*+l)+b(s^*+l)^2-(al+bl^2)), \tag{2.15}$$

holds for all $\tau_0 \leq T$, where $c \geq 1$. This is because the integral over the entire time T
on both sides leads to

$$\int_0^T (a(\hat{s}+l)+b(\hat{s}+l)^2-(al+bl^2))\mathrm{d}t$$
$$\leq c \cdot \int_0^T (a(s^*+l)+b(s^*+l)^2-(al+bl^2))\mathrm{d}t, \tag{2.16}$$

where $\int_0^T (a(\hat{s}+l)+b(\hat{s}+l)^2)\mathrm{d}t$ is the total cost of ORCHARD, $\int_0^T (a(s^*+l)+b(s^*+l)^2)\mathrm{d}t$ is the cost of optimal offline algorithm. In this sense, (2.16) is consistent
with the definition of competitive ratio in (2.9). Before providing the proof of
competitiveness, we introduce the following notations. At a current time τ_0 in the
ORCHARD algorithm, let $\hat{w}(t',t''), \tau_0 \leq t' \leq t''$ denote the total residual demand of
PEVs whose deadlines are between $[t',t'']$. Similarly, for offline optimal algorithm,
let $w^*(t',t''), \tau_0 \leq t' \leq t''$ denote the total residual demand of PEVs whose deadlines

are between $[t', t'']$. Note that $\hat{w}(t', t'')$ and $\tilde{w}(t', t'')$ are likely to be rather different since the charging solution before current time τ_0 of ORCHARD and offline optimal algorithm are very likely different. We further denote

$$
\begin{aligned}
d(t', t'') = \max \Big\{ &0, \min\{\hat{w}(t', t''), \frac{1}{q} \sum_{i \in \mathscr{I}(t')} U_i(t'' - t')\} \\
&- \min\{w^*(t', t''), \sum_{i \in \mathscr{I}(t')} U_i(t'' - t')\} \Big\}
\end{aligned}
\tag{2.17}
$$

as the amount of additional demand left for ORCHARD with deadline in $(t', t'']$. Then, we define a sequence of time points $\tau_1, \tau_2, \cdots$ as follows: let τ_1 be the time such that $d(\tau_0, \tau_1)/(\tau_1 - \tau_0)$ is maximized. If there are several such points, we choose the furthest one. Given τ_k, we let $\tau_{k+1} > \tau_k$ be the furthest point that maximizes $d(\tau_k, \tau_{k+1})/(\tau_{k+1} - \tau_k)$, i.e.,

$$
\tau_{k+1} = \arg \max_{\tau > \tau_k} d(\tau_k, \tau)/(\tau - \tau_k).
\tag{2.18}
$$

The "load intensity gap" within $(\tau_k, \tau_{k+1}]$ is denoted as

$$
g_k = d(\tau_k, \tau_{k+1})/(\tau_{k+1} - \tau_k), k = 1, 2, \cdots.
\tag{2.19}
$$

Evidently, g_k is a non-negative monotonically decreasing sequence.

We are now ready to define the potential function Φ as

$$
\Phi = \beta_1 \cdot a \sum_{k=0}^{\infty} ((\tau_{k+1} - \tau_k)g_k) + \beta_2 \cdot b \sum_{k=0}^{\infty} ((\tau_{k+1} - \tau_k)g_k^2),
\tag{2.20}
$$

where β_1, β_2 will be assigned finite values later. We notice that $\Phi(0) = \Phi(T) = 0$ holds, since the load is clearly zero before any PEV arrives and after the last deadline.

In the Theorem 2.1 below, we derive the competitive ratio of ORCHARD. First, we provide the following Lemma to be used in proving Theorem 2.1.

Lemma 2.3.

$$
qg_0 \leq \hat{s} \leq qg_0 + qs^*.
\tag{2.21}
$$

The detailed proof please see Appendix "Proof of Lemma 2.3".

In the Theorem 2.1 below, we derive the competitive ratio of ORCHARD.

Theorem 2.1. *ORCHARD is 2.39-competitive when the cost function is a quadratic function of the load demand by setting $q = 1.46$.*

Proof. Please see the detailed proof in Appendix "Proof of Theorem 2.1".

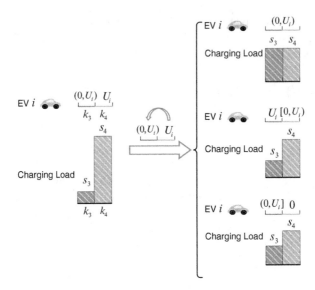

Fig. 2.4 Illustration of example based on KKT conditions. In case 1, we have $y^*_{k_1} = y^*_{k_2}$. In case 2 and 3, we cannot achieve the balanced charging rate, i.e., $y^*_{k_1} < y^*_{k_2}$, either because $x^*_{ik_1}$ has increased to the limit U_i (case 2) or $x^*_{ik_2}$ has decreased to 0 (case 3)

2.4 The Optimal Offline Algorithm with Low Complexity

The major complexity of Algorithm 1 lies in the computation involved in solving Problem (2.12) every time when a PEV arrives or finishes charging. By exploring the special structure of the optimal solution, we propose in this section a low-complexity solution algorithm to solve problem (2.12). Notice that Problem (2.12) and the offline optimization problem (2.8) have exactly the same structure. Both of them are to minimize a convex and additive objective function over a polyhedron. Thus, the algorithm proposed here can also apply to (2.12). The proposed algorithm is shown to have a much lower computational complexity than generic convex optimization algorithms, such as the interior point method.

2.4.1 Optimality Analysis

It is easy to verify that problem (2.8) is convex, and then we apply the Karush-Kuhn-Tucker (KKT) conditions to it [17]. We associate a dual variable λ_i with inequality (2.8a), a dual variable w_{ik} with inequality (2.8b), a dual variable v_{ik} with inequality (2.8c). Then, the Lagrangian is given by:

$$L = \sum_{k \in \mathcal{K}} (a(\sum_{j \in \mathcal{I}(k)} x_{jk} + l_k) + b(\sum_{j \in \mathcal{I}(k)} x_{jk} + l_k)^2$$

$$- (al_k + bl_k^2))\delta_k - \sum_{i=1}^{N} \lambda_i (\sum_{k \in \mathcal{J}(i)} x_{ik}\delta_k - D_i) \qquad (2.22)$$

$$+ \sum_{i=1}^{N} \sum_{k \in \mathcal{J}(i)} (-\omega_{ik}x_{ik} + v_{ik}(x_{ik} - U_i)).$$

Let x_{ik}^* denote the optimal charging rate for EV i at interval $k \in \mathcal{I}(k)$. The necessary and sufficient *KKT* conditions are given by:

$$a + 2b(\sum_{j \in \mathcal{I}(k)} x_{jk}^* + l_k) - \lambda_i + v_{ik} - \omega_{ik} = 0,$$

$$i = 1,\ldots,N, k \in \mathcal{J}(i), \qquad (2.23a)$$

$$\lambda_i(D_i - \sum_{k \in \mathcal{J}(i)} x_{ik}^*) = 0, i = 1,\ldots,N, \qquad (2.23b)$$

$$\omega_{ik}x_{ik}^* = 0, i = 1,\ldots,N, k \in \mathcal{J}(i), \qquad (2.23c)$$

$$v_{ik}(x_{ik}^* - U_i) = 0, i = 1,\ldots,N, k \in \mathcal{J}(i), \qquad (2.23d)$$

where (2.23a) means that the differentiation of L should be 0 at x_{ik}^*, and (2.23b), (2.23c), (2.23d) are the complementary slackness conditions. We separate our analysis into the following three cases:

1. If $x_{ik_1}^* \in (0, U_i)$ for a particular PEV i in a time interval $k_1 \in \mathcal{J}(i)$, then, by complementary slackness, we have $v_{ik_1} = w_{ik_1} = 0$. From (2.23a), $y_{k_1} = \sum_{j \in \mathcal{I}(k_1)} x_{jk_1} + l_{k_1} = (\lambda_i - a)/2b$.
2. If $x_{ik_2}^* = 0$ for PEV i during a time interval $k_2 \in \mathcal{J}(i)$, we can infer from (2.23c) and (2.23d) that $\omega_{ik_2} > 0$ and $v_{ik_2} = 0$. Then, $y_{k_2} = \sum_{j \in \mathcal{I}(k_2)} x_{jk_2}^* + l_{k_2} = (\lambda_i - a)/2b + \omega_{ik_2}/2b$.
3. Similarly, if $x_{ik_3}^* = U_i$ for PEV i in interval $k_3 \in \mathcal{J}(i)$, then, we have $y_{k_3} = \sum_{j \in \mathcal{I}(k_3)} x_{jk_3}^* + l_{k_3} = (\lambda_i - a)/2b - v_{ik_3}/2b$.

Let y_k^* be the optimal total load that

$$y_k^* = \sum_{i \in \mathcal{I}(k)} x_{ik}^* + l_k. \qquad (2.24)$$

From the above discussions, we can conclude that the necessary and sufficient conditions for the optimal total charging rate as follows:

1. y_k^* is the same for a set of intervals as long as there exists a PEV i that parks through this set of intervals with $x_{ik}^* \in (0, U_i)$.

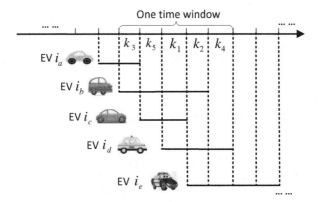

Fig. 2.5 Illustration of one time window. The time window starts from the arrival time of PEV i_b to the departure time of PEV i_d

2. If $x^*_{ik} = 0$ for a PEV i during an interval k that it parks in, then, y^*_k in that interval is no smaller than that of the other interval $k' \in \mathcal{J}(i)$ during which $x^*_{ik'} \in (0, U_i]$.
3. If $x^*_{ik} = U_i$ for PEV i during an interval k, then, y^*_k is no larger than that of the other interval $k' \in \mathcal{J}(i)$ whose charging rate $x^*_{ik'} \in [0, U_i)$.

The above conditions can be intuitively understood as follows. Due to the convexity the objective function, the optimal solution to (2.8) always tries to balance the total load y_k across different intervals. For example, as shown in Fig. 2.4, if there are two intervals k_1 and k_2 with $y^*_{k_1} < y^*_{k_2}$, and a PEV i such that $x^*_{ik_1} = 0$ and $0 < x^*_{ik_2} \leq U_i$, then we can always shift the charging load of PEV i from interval k_2 to k_1 to decrease the total load difference between k_1 and k_2. In other words, whenever possible, the charging load should be shifted from interval k_2 to k_1 until $y^*_{k_1} = y^*_{k_2}$ (case 1 in Fig. 2.4). However, such balanced charging rate at two intervals may not achievable, resulting $y^*_{k_1} < y^*_{k_2}$, either because $x^*_{ik_1}$ has increased to the limit U_i (case 2) or $x^*_{ik_2}$ has decreased to 0 (case 3). Based on these conditions, we present a low-complexity solution algorithm in the next subsection.

2.4.2   Algorithm Description

From the analysis of KKT optimality conditions, one should manage to balance the total load among all intervals under the constraints of each individual PEV's charging profiles. In this subsection, we present a charging rate allocation algorithm to achieve the objective of "load balancing". The optimality and complexity of the proposed algorithm will be discussed in the next subsection.

Intuitively, one should shift the demand from "heavily loaded" intervals to the others. To do this, we first introduce the concept of *intensity* of an interval k, denoted by ρ_k, to quantify the heaviness of the load in the interval. Specifically, ρ_k is defined as the upper bound of the charging load of an interval, and is given by:

$$\rho_k = \sum_{i \in \mathscr{I}(k)} \min\left\{U_i, \frac{D_i}{\delta_k}\right\}. \tag{2.25}$$

This is because the charging rate of each PEV i in the interval k will not exceed the minimum between the charging rate bound U_i and the D_i/δ_k, i.e., PEV i only charges in the interval k. The basic idea of the proposed algorithm is to shift the demand of a set of intervals with high intensities to the others with lower intensities. Notice that the demand of an interval k_1 can only be transferred to its neighboring interval k_2 such that $k_1 \in J(i)$ and $k_2 \in J(i)$ hold for some PEV i. Therefore, we need to consider both the intensities of an interval set and their neighboring intervals to make the decision on "load balancing".

From the above discussion, we take into consideration a set of consecutive intervals, referred to as a "time window", starting from the arrival time of a PEV to the departure time of one, probably another PEV. If there are N PEVs, the maximum number of time windows is N^2. Within a tagged time window, we select a set of intervals of the highest intensities as the candidate interval set from which the load is to be transferred to the other intervals in the time window. In practice, we first consider the single interval with the highest intensity, then the top two intervals, top three intervals, etc. That is, for each time window, sort the intervals in descending order according to ρ. The index is denoted by $k_1, k_2, \ldots$, as illustrated in Fig. 2.5.

Evidently, a time window consisting of K' intervals contains K' such interval sets. For example, there are 5 interval sets in the time window shown in Fig. 2.5. We denote the interval sets obtained from all the time windows in the entire duration T as $\mathscr{K}_1, \mathscr{K}_2, \cdots$. Then, the following iterative algorithm determines the load transfer operation of intervals as well as the charging rate schedule of all PEVs.

Step 1: For each interval set $\mathscr{K}$, we first compute the residual demand of PEV i on $\mathscr{K}$. The residual demand of PEV i on $\mathscr{K}$, denoted by $D_i(\mathscr{K})$, is calculated by letting PEV i be charged at the upper bound U_i on its parking intervals non-overlapped with $\mathscr{K}$. That is

$$D_i(\mathscr{K}) = D_i - U_i \sum_{k \in \mathscr{J}(i) \backslash (\mathscr{J}(i) \cap \mathscr{K})} \delta_k. \tag{2.26}$$

The intuition is to transfer as much as possible the charging demand from intervals with high intensities to its neighboring intervals. Then, we can calculate the total load of the interval set $\mathscr{K}$ by balancing the residual demand over all the intervals in $\mathscr{K}$, i.e.,

$$y = \frac{\sum_{k \in \mathscr{K}} \left(\sum_{i \in \mathscr{I}(k)} (\max\{0, D_i(\mathscr{K})\}) + \hat{y}_k \delta_k\right)}{\sum_{k \in \mathscr{K}} \delta_k}, \tag{2.27}$$

where $\hat{y}_k$ is the total load after scheduling in previous iterations at the interval and initially set to be l_k.

Step 2: Find the interval set $\mathcal{K}^*$ with the highest total load y^*. Then the optimal total charging rate of interval in $\mathcal{K}^*$ is set to be s_k^*, where

$$s_k^* = y^* - l_k, \forall k \in \mathcal{K}^*. \tag{2.28}$$

We denote $\mathcal{I}^*$ by the set of PEVs of which the residual demand $D_i(\mathcal{K}^*)$ is non-negative, Δ^* by the total length of the intervals in the set $\mathcal{K}^*$, i.e., $\Delta^* = \sum_{k \in \mathcal{K}^*} \delta_k$. For each PEV $i \in \mathcal{I}^*$, we schedule the charging rate as

$$x_{ik}^* = \begin{cases} U_i - \frac{(U_i\Delta^* - D_i(\mathcal{K}^*))(\sum_i U_i - s_k^*)}{\sum_i (U_i\Delta^* - D_i(\mathcal{K}^*))}, & k \in \mathcal{K}^*, \\ U_i, & k \in \mathcal{J}(i) \setminus \mathcal{K}^*. \end{cases} \tag{2.29}$$

It is easy to verify that $\sum_{k \in \mathcal{K}^*} x_{ik}^* = s_k^*$ for $k \in \mathcal{K}^*$. Note that PEV $i \in \mathcal{I}^*$ has finished scheduled charging rate and will not be considered in the next iterations. Then the total charging rate at any interval $k \in \mathcal{J}(i) \setminus \mathcal{K}^*$ should be increased by U_i. We use $\hat{s}_k$ to denote the total rate scheduled in the interval $k \notin \mathcal{K}^*$ up to the current iteration, which is updated as.

$$\hat{s}_k = \hat{s}_k + \sum_{i \in \mathcal{I}^* \cap \mathcal{J}(k)} U_i. \tag{2.30}$$

For a PEV $i \notin I^*$ whose parking intervals overlaps with $\mathcal{K}^*$, the charging rate of its parking intervals overlapped with $\mathcal{K}^*$ is assigned to be 0, i.e.,

$$x_{ik}^* = 0, k \in \mathcal{J}(i) \cap \mathcal{K}^*. \tag{2.31}$$

Step 3: Exclude $\mathcal{I}^*$ and $\mathcal{K}^*$ from the PEV set and interval set, and merge the remaining intervals into a new time duration. Find all the interval sets in the newly formed time windows as in Fig. 2.5. Then, repeat from step 1 until the charging rates of all PEVs are scheduled.

2.4.3 Optimality and Complexity

We first provide the following Lemma 2.4 before proving the global optimality of the proposed algorithm. Denote $\mathcal{K}^*(m)$ by the interval set found in mth iteration, $\Delta^*(m)$ by the total length of intervals in $\mathcal{K}^*(m)$, i.e,

$$\Delta^*(m) = \sum_{k \in \mathcal{K}^*(m)} \delta_k, \tag{2.32}$$

$y^*(m)$ by the highest total load of interval set $\mathcal{K}^*(m)$ respectively.

Lemma 2.4. *In the proposed low-complexity solution algorithm, the highest total load found in mth iteration is no smaller than that found in $(m+1)$th iteration, i.e., $y^*(m) \geq y^*(m+1)$.*

Please see the proof in Appendix "Proof of Lemma 2.4".

Theorem 2.2. *The proposed algorithm always outputs a globally optimal schedule.*

Proof. Please see the detailed proof in Appendix "Proof of Theorem 2.2".

Now we give a complexity analysis of the proposed algorithm. Consider the worst case where N PEVs lead to $2N-1$ intervals, N^2 variables and $2N^2 + N$ constrains. The proposed low-complexity solution algorithm at least excludes one interval in each outer loop that leads to at most $2N-1$ iterations. In each iteration (step 1 - step 3), there are at most $N(N+1)/2$ time windows which contains at most N possible interval sets. Hence, the total number of iterations is in the order of $O(N^4)$. Since the operation complexity of intensity calculation for each sequence is $O(N)$ (we regard one addition, subtraction, multiplication and division as one operation), the upper bound of operation complexity is $O(N^5)$. On the other hand, the generic interior point algorithm has a complexity at the order of $O(n^{3.5})$ [18], where n is the number of variables. Note that $n = N^2$ in our problem, and thus the complexity of interior point algorithm is $O(N^7)$, which is much higher than that of the proposed algorithm.

2.5 Simulations

In this section, we evaluate the performance of ORCHARD and verify the iteration complexity of the low-complexity solution algorithm. Specially, we define the *average performance ratio* as the ratio of the average cost of online algorithm to that of offline optimal algorithm. Note that the variation of a and b, $(a, b > 0)$ will not change charging solution to ORCHARD while the variation of q will, so we only discuss how the average performance ratio changes by varying q, shown in Sect. 2.5.3.

2.5.1 Performance Ratio Evaluation

We consider a running time T of 24 hours. We choose the base load profile of one day in the service area of South California Edison from [4]. The coefficients of the cost function are set to $a = 10^{-4}$ \$/kWh and $b = 0.6 \times 10^{-4}$ \$/kWh/kW [3]. There are two types of PEVs in our simulation [19]: (1) maximum charging rate $U_i = 3.3$ kW, battery capacity $\zeta_i = 35$ kWh; (2) maximum charging rate $U_i = 1.4$ kW, battery capacity $\zeta_i = 16$ kWh. Each PEV is equally likely chosen from the two

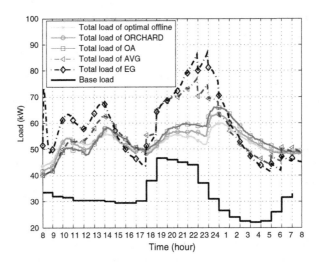

Fig. 2.6 Base load and total load of five algorithms

Table 2.1 Parameter settings of the arrival and parking durations

Time of day	Arrival rate (PEVs/hour)	Mean parking Time (hour)
08:00–10:00	7	10
10:00–12:00	5	1/2
12:00–14:00	10	2
14:00–18:00	5	1/2
18:00–20:00	10	2
20:00–24:00	5	10
24:00–08:00	0	0

types and the charging demand is uniformly chosen from $[0, \min\{U_i \cdot (t_i^{(e)} - t_i^{(s)}), \zeta_i\}]$ (this ensures that (2.8) is feasible). Each PEV's arrival follows a Poisson distribution and the parking time follows an exponential distribution [9]. The mean arrival and parking durations are listed in Table 2.1, where there are three peak hours with large arrival rates, i.e. 8 to 10, $12:00$ to $14:00$ and $18:00$ to $20:00$. The settings of the peak hour match with the realistic vehicle trips in National Household Travel Survey (NHTS) 2009 [20].

We compare ORCHARD to the optimal offline algorithm as well as other online algorithms. Unless otherwise specified, the speeding factor of ORCHARD, q, is set to be 1.46. Note that by Theorem 2.1 $q = 1.46$ achieves the best competitive ratio in the worst case, but may not be the best choice for average performance. We will discuss the effect of q in Sect. 2.5.3. We denote the cost of ORCHARD and the optimal offline algorithm by Ψ_{ORC} and Ψ^*, respectively. The other online algorithms for comparison are

Table 2.2 Average performance ratio of online algorithms

$\frac{\Psi_{ORC}}{\Psi^*}$	$\frac{\Psi_{OA}}{\Psi^*}$	$\frac{\Psi_{AVG}}{\Psi^*}$	$\frac{\Psi_{EG}}{\Psi^*}$
1.065	1.134	1.528	2.344

Table 2.3 Parameter settings of the three scenarios

Time of day	Arrival rate (PEVs/hour)			Mean parking
	S. 1	S. 2	S. 3	Time (hour)
08:00–10:00	7	7	7	10
10:00–12:00	5	5	5	1/2
12:00–14:00	10	30	50	2
14:00–18:00	5	5	5	1/2
18:00–20:00	10	30	50	2
20:00–24:00	5	5	5	10
24:00–08:00	0	0	0	0

1. online average charging (AVG): The charging demand is evenly distributed during the parking period, i.e. the charging rate is $D_i/(t_i^{(e)} - t_i^{(s)})$.
2. online eagerly charging (EG): PEV i is charged at the maximum charging rate U_i.
3. online optimal available information charging (OA) : Set $q = 1$ in ORCHARD.

Their costs are denoted by Ψ_{AVG}, Ψ_{EG} and Ψ_{OA}, respectively.

All the convex optimizations are solved by CVX [21]. We simulate 10^5 cases and plot the average base load as well as the total load over time in Fig. 2.6, where the total load represents the sum of the base load and the charging load, defined in Eq. (2.3). In addition, the average performance ratios normalized against the optimal offline solution are shown in Table 2.2. Figure 2.6 shows that the total load curve of ORCHARD follows closely with the optimal offline solution curve. In contrast, EG and AVG largely deviate from the optimal charging curve, being either too aggressive or too conservative depending on the arrival patterns. From Table 2.2, we can see that ORCHARD performs the best among the four online algorithms, which has on average less than 6.5 % extra cost compared with the optimal offline algorithm.

2.5.2 The Influence by the PEV Pattern

In this subsection, we ignore the base load that mainly discuss how PEV pattern affects the average performance ratio. We consider three different scenarios, whose mean arrival and parking durations are listed in Table 2.3. In particular, scenarios 1–3 represent light, moderate and heavy traffic, respectively. The main difference lies in the arrival rates at the two peak hours, i.e. $12:00$ to $14:00$ and $18:00$ to $20:00$.

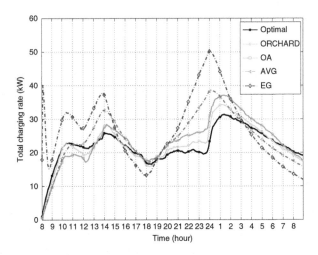

Fig. 2.7 PEV total charging rate of five algorithms in Scenario 1

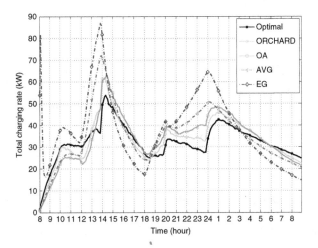

Fig. 2.8 Base load and total load of five algorithms in Scenario 2

For each scenario, we simulate 10^5 cases and plot the average total charging rate over time in Figs. 2.7, 2.8 and 2.9, respectively, where the vertical axis is the total charging rate of PEVs, defined in Eq. (2.2). In addition, the average performance ratios normalized against the optimal offline solution are shown in Table 2.4. In all scenarios, ORCHARD performs the best among the four online algorithms, which has on average less than 14% extra cost compared with the optimal offline algorithm. We also notice that ORCHARD has a 10% performance gain compared with the OA algorithm in the scenario with heavy traffic. We will discuss the proper setting of q in Sect. 2.5.3. The charging rate curve of the proposed online charging algorithm follows closely with the optimal offline solution curve. In

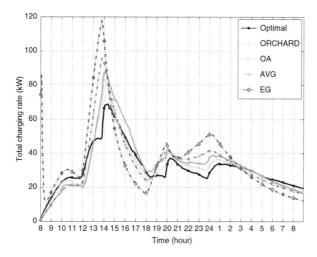

Fig. 2.9 Base load and total load of five algorithms in Scenario 3

Table 2.4 Average performance ratio of online algorithms	Scenario	$\frac{\Psi_{ORC}}{\Psi^*}$	$\frac{\Psi_{OA}}{\Psi^*}$	$\frac{\Psi_{AVG}}{\Psi^*}$	$\frac{\Psi_{EG}}{\Psi^*}$
	1	1.068	1.135	1.530	2.346
	2	1.104	1.197	1.645	2.309
	3	1.133	1.240	1.701	2.273

contrast, EG and AVG largely deviate from the optimal charging curve, being either too aggressive or too conservative depending on the arrival patterns. In general, all charging algorithms perform better when the traffic is relatively light, except for EG. It produces even the worst performance ratio under light traffic. This is partly because its aggressive charging scheme somehow matches with the large traffic variations in scenario 3.

2.5.3 Setting a Proper q

Theoretically, setting q to be 1.46 will achieve the best ratio in the worst case. However, it does not achieve the best average performance in general. In this subsection, we discuss how q affects the normalized average performance ratio. For the three scenarios with different traffic, we plot the normalized average performance ratio in Fig. 2.10 by varying q from 1 to 5. For scenario 1, setting $q = 1.8$, $\frac{\Psi_{ORC}}{\Psi^*}$ achieves the lowest average ratio 1.053. For scenario 2, setting $q = 2.1$, $\frac{\Psi_{ORC}}{\Psi^*}$ achieves the lowest average ratio 1.052. For scenario 3, setting $q = 2.3$, $\frac{\Psi_{ORC}}{\Psi^*}$ achieves the lowest average ratio 1.050, which is about 8% lower than that when $q = 1.46$. In general, the optimal q is larger when the traffic is heavy and unpredictable as in scenario 3. Intuitively, this is because the energy cost during

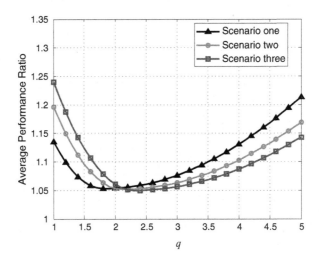

Fig. 2.10 Average performance ratios of ORCHARD in three scenarios with varied q

peak arrivals largely dominates the overall cost. A larger q is able to better utilize off-peak hour and to speed up charging when peak hours arrive. Here, we provide a simple method to achieve a better average performance by adjusting q. In practice, a charging station can collect the past data on PEV patterns, based on which, the value of q can be searched for the best average performance.

2.5.4 Complexity of Low-Complexity Solution Algorithm

To verify the iteration complexity of the low-complexity solution algorithm, we adopt it to solve problem (2.8). For the system parameter, we use the same settings with the default settings except the arrival rates, which are assumed to be the same during $8:00 - 18:00$ and 0 after $18:00$. We vary the arrival rate in $8:00 - 18:00$ from 1 to 10 (PEVs/hour) that leads to the mean of N(the number of total PEVs one day) varies from 10 to 100. For each specified mean of N, we simulate 10000 times and compute the average number of iterations and operations. The result is shown in Fig. 2.11. We also fit the data of iterations and operations with the polynomial function $f(x) = 2019.5x^4 - 297.9x^3 + 18.6x^2 - 0.2x$ and $f(x) = 166010x^5 - 26270x^4 + 1200x^3 - 10x^2$ with the *mean relative error* 0.039 and 0.054 respectively. It shows that both the iteration and operation complexity match our complexity analysis in Section IV.C.

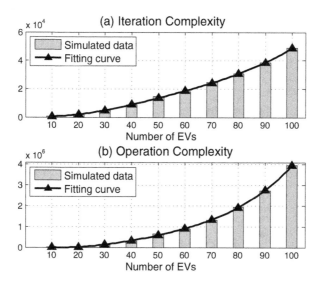

Fig. 2.11 Iterations and operations of low complexity algorithm

2.6 Conclusions

In this chapter, we have proposed an Online cooRdinated CHARging Decision (ORCHARD) algorithm, which minimizes the energy cost and flattenes the total electric load profile without knowing the future information. Through rigorous proof, we showed that ORCHARD is strictly feasible in the sense that it guarantees to fulfill all charging demands before due time. Meanwhile, it achieves the best known competitive ratio of 2.39 when the cost function is a quadratic function of the load demand. To further reduce the computational complexity of the algorithm, we proposed a novel reduced-complexity algorithm to replace the standard convex optimization techniques used in ORCHARD. Through extensive simulations, we showed that the average performance gap between ORCHARD and the optimal offline solution, which utilizes the complete future information, is as small as 6.5 %. By setting proper speeding factor, the average performance gap can be further reduced to less than 5 %.

Appendix

Proof of Lemma 2.1

Proof. The proof is given by contradiction. The optimal total charging rate at time $t \in [t_k, t_{k+1})$ is denoted by $\tilde{s}_k(t)$, where

$$\tilde{s}_k(t) = \sum_{i \in I(t_k)} x_{it}^*, k = 1, \ldots, K. \tag{2.33}$$

Let

$$s_k = \frac{\int_{t_k}^{t_{k+1}} \tilde{s}_k(t) dt}{t_{k+1} - t_k} \tag{2.34}$$

be the average charging rate in δ_k. Note that s_k is always achievable by setting the charging rate of each EV i as $\int_{t_k}^{t_{k+1}} x_{it}^* dt / (t_{k+1} - t_k)$. By Jensen's inequality, we have

$$\int_{t_k}^{t_{k+1}} \frac{a(\hat{s}_k(t) + l_k) + b(\hat{s}_k(t) + l_k)^2}{t_{k+1} - t_k} dt$$

$$\geq a \frac{\int_{t_k}^{t_{k+1}} \hat{s}_k(t) + l_k dt}{t_{k+1} - t_k} + b \left[\frac{\int_{t_k}^{t_{k+1}} \hat{s}_k(t) + l_k dt}{t_{k+1} - t_k} \right]^2 \tag{2.35}$$

$$= a(s_k + l_k) + b(s_k + l_k)^2.$$

Equivalently, we have

$$\int_{t_k}^{t_{k+1}} \left[a(\hat{s}_k(t) + l_k) + b(\hat{s}_k(t) + l_k)^2 - (al_t + bl_t^2) \right] dt$$

$$\geq (t_{k+1} - t_k) (a(s_k + l_k) + b(s_k + l_k)^2 - (al_k + bl_k^2)). \tag{2.36}$$

From (2.36), the uniform total charging rate s_k incurs no higher cost than that of x_{it}^*, which contradicts with the assumption that x_{it}^* is the optimal charging schedule. Therefore, the optimal charging schedule must produce constant total charging rate in each interval δ_k, which completes the proof. ∎

Proof of Lemma 2.2

To see this, the charging rate of EV i is

$$\hat{x}_{it} = \min\{\bar{x}_{i1}(t_j) + \frac{U_i - \bar{x}_{i1}(t_j)}{\sum_{i \in \mathscr{I}_{t_j}} (U_i - \bar{x}_{i1}(t_j))} \cdot \frac{q-1}{q} \hat{s}_t, U_i\}. \tag{2.37}$$

From step 5 of ORCHARD, the total charing rate is

$$\hat{s}_t = q \cdot \sum_{i \in \mathscr{I}_{t_j}} \bar{x}_{i1}(t_j), \tag{2.38}$$

if

$$q \cdot \sum_{i \in \mathscr{I}_{t_j}} \bar{x}_{i1}(t_j) \leq \sum_{i \in \mathscr{I}_{t_j}} U_i. \tag{2.39}$$

In this case, the charging rate of PEV i is

$$\bar{x}_{i1}(t_j) + \frac{U_i - \bar{x}_{i1}(t_j)}{\sum_{i \in \bar{\mathscr{I}}_{t_j}} (U_i - \bar{x}_{i1}(t_j))} \frac{q-1}{q} \hat{s}_t$$

$$= \bar{x}_{i1}(t_j) + \frac{U_i - \bar{x}_{i1}(t_j)}{\sum_{i \in \bar{\mathscr{I}}_{t_j}} (U_i - \bar{x}_{i1}(t_j))} \cdot (q-1) \sum_{i \in \bar{\mathscr{I}}_{t_j}} \bar{x}_{i1}(t_j) \tag{2.40}$$

$$\leq \bar{x}_{i1}(t_j) + \frac{U_i - \bar{x}_{i1}(t_j)}{\sum_{i \in \bar{\mathscr{I}}_{t_j}} (U_i - \bar{x}_{i1}(t_j))} \cdot \left(\sum_{i \in \bar{\mathscr{I}}_{t_j}} U_i - \sum_{i \in \bar{\mathscr{I}}_{t_j}} \bar{x}_{i1}(t_j) \right)$$

$$= U_i.$$

Otherwise if (2.39) does not hold, the total online charging rate is

$$\hat{s}_t = \sum_{i \in \bar{\mathscr{I}}_{t_j}} U_i. \tag{2.41}$$

Then, we have

$$\bar{x}_{i1}(t_j) + \frac{U_i - \bar{x}_{i1}(t_j)}{\sum_{i \in \bar{\mathscr{I}}_{t_j}} (U_i - \bar{x}_{i1}(t_j))} \cdot \frac{q-1}{q} \hat{s}_t. \tag{2.42}$$

By step 4, $\hat{x}_{it} = U_i$, and also it is easy to verify that $\sum_{i \in \bar{\mathscr{I}}_{t_j}} \hat{x}_{it} = \hat{s}_t$. To sum up, in any case, the following constraints are satisfied, i.e.

$$\bar{x}_{i1}(t_j) \leq \hat{x}_{it} \leq U_i, \quad \sum_{i \in \bar{\mathscr{I}}_{t_j}} \hat{x}_{it} = \hat{s}_t. \tag{2.43}$$

On the other hand, the charging schedule $\hat{x}_{it}$ can finish the charging of all EV's before their departures. This is because it is no slower than the optimal charging schedule $\bar{x}_{i1}(t_j)$, which guarantees the feasibility of (2.7). ∎

Proof of Lemma 2.3

Based on the definition, we have following two inequalities

$$\frac{\hat{w}(\tau_0, \tau_1)}{\tau_1 - \tau_0} \leq \sum_{i \in \mathscr{I}(\tau_0)} U_i, \tag{2.44a}$$

$$\frac{w^*(\tau_0, \tau_1)}{\tau_1 - \tau_0} \leq \sum_{i \in \mathscr{I}(\tau_0)} U_i, \tag{2.44b}$$

which hold because all the PEVs with deadlines in $[\tau_0, \tau_1]$ must park in the station at current time τ_0 such that $\sum_{i \in \mathscr{I}(\tau_0)} U_i$ is larger or equal to $\sum_{i \in \mathscr{I}(t)} U_i$ for $t \in (\tau_0, \tau_1]$. Due to the setting of $\hat{s}$ in our online algorithm, either the inequality

$$q\frac{\hat{w}(\tau_0, \tau_1)}{\tau_1 - \tau_0} \leq \hat{s} < \sum_{i \in \mathscr{I}(\tau_0)} U_i \tag{2.45}$$

or

$$\hat{s} = \sum_{i \in \mathscr{I}(\tau_0)} U_i \leq q\frac{\hat{w}(\tau_0, \tau_1)}{\tau_1 - \tau_0} \tag{2.46}$$

holds. Similarly, for optimal total charging rate s^* in offline algorithm, the inequality

$$\frac{w^*(\tau_0, \tau_1)}{\tau_1 - \tau_0} \leq s^* \leq \sum_{i \in \mathscr{I}(\tau_0)} U_i \tag{2.47}$$

holds since $w^*(\tau_0, t)$ does not include the demand of the future coming PEVs while s^* dose. From the definition of g_k, we get that

$$g_0 = \max\left\{0, \min\{\frac{\hat{w}(\tau_0, \tau_1)}{\tau_1 - \tau_0}, \frac{1}{q} \sum_{i \in \mathscr{I}(\tau_0)} U_i\} - \right.$$
$$\left. \min\{\frac{w^*(\tau_0, \tau_1)}{\tau_1 - \tau_0}, \sum_{i \in \mathscr{I}(\tau_0)} U_i\}\right\} \tag{2.48}$$

To further reduce g_0, we need to discuss the following four cases.

Case 1: If

$$q\frac{\hat{w}(\tau_0, \tau_1)}{\tau_1 - \tau_0} \geq \sum_{i \in \mathscr{I}(\tau_0)} U_i \text{ and } \frac{w^*(\tau_0, \tau_1)}{\tau_1 - \tau_0} = \sum_{i \in \mathscr{I}(\tau_0)} U_i, \tag{2.49}$$

then from (2.46), (2.47) we get

$$\hat{s} = s^* = \sum_{i \in \mathscr{I}(\tau_0)} U_i \tag{2.50}$$

and

$$g_0 = \max\left\{0, \frac{1}{q}\sum_{i\in\mathscr{I}(\tau_0)} U_i - \sum_{i\in\mathscr{I}(\tau_0)} U_i\right\} = 0. \tag{2.51}$$

Hence,

$$qg_0 = 0 \le \hat{s} = \sum_{i\in\mathscr{I}(\tau_0)} U_i \le q\sum_{i\in\mathscr{I}(\tau_0)} U_i = qg_0 + qs^*. \tag{2.52}$$

Case 2: If

$$q\frac{\hat{w}(\tau_0,\tau_1)}{\tau_1 - \tau_0} < \sum_{i\in\mathscr{I}(\tau_0)} U_i \text{ and } \frac{w^*(\tau_0,\tau_1)}{\tau_1 - \tau_0} = \sum_{i\in\mathscr{I}(\tau_0)} U_i, \tag{2.53}$$

then from (2.45), (2.47) we get

$$\hat{s} \le s^* = \sum_{i\in\mathscr{I}(\tau_0)} U_i \tag{2.54}$$

and

$$g_0 = \max\left\{0, \frac{\hat{w}(\tau_0,\tau_1)}{\tau_1 - \tau_0} - \sum_{i\in\mathscr{I}(\tau_0)} U_i\right\} = 0. \tag{2.55}$$

Hence,

$$qg_0 = 0 \le \hat{s} \le q\sum_{i\in\mathscr{I}(\tau_0)} U_i = qg_0 + qs^*. \tag{2.56}$$

Case 3: If

$$q\frac{\hat{w}(\tau_0,\tau_1)}{\tau_1 - \tau_0} \ge \sum_{i\in\mathscr{I}(\tau_0)} U_i \text{ and } \frac{w^*(\tau_0,\tau_1)}{\tau_1 - \tau_0} < \sum_{i\in\mathscr{I}(\tau_0)} U_i, \tag{2.57}$$

then from (2.46), (2.47) we get

$$s^* \le \hat{s} = \sum_{i\in\mathscr{I}(\tau_0)} U_i \tag{2.58}$$

and

$$g_0 = \max\left\{0, \frac{1}{q}\sum_{i\in\mathscr{I}(\tau_0)} U_i - \frac{w^*(\tau_0,\tau_1)}{\tau_1 - \tau_0}\right\}. \tag{2.59}$$

If

$$\frac{1}{q} \sum_{i \in \mathscr{I}(\tau_0)} U_i \le \frac{w^*(\tau_0, \tau_1)}{\tau_1 - \tau_0}, \tag{2.60}$$

then we get $g_0 = 0$, and

$$q g_0 = 0 \le \hat{s} = \sum_{i \in \mathscr{I}(\tau_0)} U_i \le q \frac{w^*(\tau_0, \tau_1)}{\tau_1 - \tau_0} \le q s^* = q g_0 + q s^*. \tag{2.61}$$

If

$$\frac{1}{q} \sum_{i \in \mathscr{I}(\tau_0)} U_i > \frac{w^*(\tau_0, \tau_1)}{\tau_1 - \tau_0}, \tag{2.62}$$

then we get

$$g_0 = \frac{1}{q} \sum_{i \in \mathscr{I}(\tau_0)} U_i - \frac{w^*(\tau_0, \tau_1)}{\tau_1 - \tau_0}. \tag{2.63}$$

Hence, from (2.58) we have

$$q g_0 = \sum_{i \in \mathscr{I}(\tau_0)} U_i - q \frac{w^*(\tau_0, \tau_1)}{\tau_1 - \tau_0} \le \hat{s}, \tag{2.64a}$$

and from (2.47) we have

$$q g_0 + q s^* = \sum_{i \in \mathscr{I}(\tau_0)} U_i - q \frac{w^*(\tau_0, \tau_1)}{\tau_1 - \tau_0} + q s^* \ge \sum_{i \in \mathscr{I}(\tau_0)} U_i = \hat{s}. \tag{2.65a}$$

Since (2.21) holds for both cases, we see that (2.21) holds in Case 3.
Case 4: If

$$q \frac{\hat{w}(\tau_0, \tau_1)}{\tau_1 - \tau_0} < \sum_{i \in \mathscr{I}(\tau_0)} U_i \text{ and } \frac{w^*(\tau_0, \tau_1)}{\tau_1 - \tau_0} < \sum_{i \in \mathscr{I}(\tau_0)} U_i, \tag{2.66}$$

we get

$$g_0 = \max \left\{ 0, \frac{\hat{w}(\tau_0, \tau_1)}{\tau_1 - \tau_0} - \frac{w^*(\tau_0, \tau_1)}{\tau_1 - \tau_0} \right\}. \tag{2.67}$$

When $\hat{w}(\tau_0, \tau_1) \ge w^*(\tau_0, \tau_1)$, (2.48) is reduced to

$$g_0 = \frac{\hat{w}(\tau_0, \tau_1)}{\tau_1 - \tau_0} - \frac{w^*(\tau_0, \tau_1)}{\tau_1 - \tau_0}. \tag{2.68}$$

Recall that $\hat{s} = qs^{OA}$ where s^{OA} is the total charging rate of OA algorithm given the current demand $\hat{w}(\tau_0, \tau_1)$, and s^* is the charging rate given $w^*(\tau_0, \tau_1)$ and possible future arrivals of other PEVs. Notice that both $\hat{w}(\tau_0, \tau_1)$ and $w^*(\tau_0, \tau_1)$ do not include the charging demand of future coming PEVs, and the difference between s^{OA} and $\hat{w}(\tau_0, \tau_1)/(\tau_1 - \tau_0)$ is only resulted from the bounds of current PEVs, while the difference between s^* and $w^*(\tau_0, \tau_1)/(\tau_1 - \tau_0)$ is due to the bounds of current PEVs as well as the possible heavy load of future coming PEVs. Therefore,

$$s^{OA} - \frac{\hat{w}(\tau_0, \tau_1)}{\tau_1 - \tau_0} \le s^* - \frac{w^*(\tau_0, \tau_1)}{\tau_1 - \tau_0} \tag{2.69}$$

As $\hat{s} = qs^{OA}$, we have

$$\hat{s} - q\frac{\hat{w}(\tau_0, \tau_1)}{\tau_1 - \tau_0} \le qs^* - q\frac{w^*(\tau_0, \tau_1)}{\tau_1 - \tau_0}. \tag{2.70}$$

Hence we get the following inequalities:

$$qg_0 = q\frac{\hat{w}(\tau_0, \tau_1)}{\tau_1 - \tau_0} - q\frac{w^*(\tau_0, \tau_1)}{\tau_1 - \tau_0} \le \hat{s}, \tag{2.71a}$$

$$qg_0 + qs^* \ge (q\frac{\hat{w}(\tau_0, \tau_1)}{\tau_1 - \tau_0} - q\frac{w^*(\tau_0, \tau_1)}{\tau_1 - \tau_0})$$

$$+ (\hat{s} - q\frac{\hat{w}(\tau_0, \tau_1)}{\tau_1 - \tau_0} + q\frac{w^*(\tau_0, \tau_1)}{\tau_1 - \tau_0}) = \hat{s}, \tag{2.71b}$$

where the last inequality of (2.71a) and the first inequality of (2.71b) are derived from (2.45) and (2.70) respectively. For the case that $\hat{w}(\tau_0, \tau_1) < w^*(\tau_0, \tau_1)$, we have $g_0 = 0$ and $\hat{s} \le qs^*$ by adding on left hand side of (2.70) $q\hat{w}(\tau_0, \tau_1)/(\tau_1 - \tau_0)$ and right hand side $qw^*(\tau_0, \tau_1)/(\tau_1 - \tau_0)$. Therefore, (2.21) holds in case 4. Finally, inequality (2.21) holds in all the four cases. This completes the proof. ∎

Proof of Theorem 2.1

Proof. We can derive from (2.20) that

$$\frac{d\Phi}{d\tau_0} = \beta_1 \cdot a \sum_{k=0}^{\infty} \frac{d\left[(\tau_{k+1} - \tau_k)g_k\right]}{d\tau_0}$$
$$+ \beta_2 \cdot b \sum_{k=0}^{\infty} \frac{d\left[(\tau_{k+1} - \tau_k)g_k^2\right]}{d\tau_0}. \tag{2.72}$$

When $\hat{w}(\tau_0, \tau_1) < w^*(\tau_0, \tau_1)$, we divide into following four cases to prove that $g_0 = 0$, where τ_1 is infinity.

1. If

$$q\frac{\hat{w}(\tau_0, \tau_1)}{\tau_1 - \tau_0} \geq \sum_{i \in \mathscr{I}(\tau_0)} U_i \text{ and } \frac{w^*(\tau_0, \tau_1)}{\tau_1 - \tau_0} = \sum_{i \in \mathscr{I}(\tau_0)} U_i, \tag{2.73}$$

then (2.51) implies that $g_0 = 0$.

2. If

$$q\frac{\hat{w}(\tau_0, \tau_1)}{\tau_1 - \tau_0} < \sum_{i \in \mathscr{I}(\tau_0)} U_i \text{ and } \frac{w^*(\tau_0, \tau_1)}{\tau_1 - \tau_0} = \sum_{i \in \mathscr{I}(\tau_0)} U_i, \tag{2.74}$$

then (2.55) implies that $g_0 = 0$.

3. If

$$q\frac{\hat{w}(\tau_0, \tau_1)}{\tau_1 - \tau_0} \geq \sum_{i \in \mathscr{I}(\tau_0)} U_i \text{ and } \frac{w^*(\tau_0, \tau_1)}{\tau_1 - \tau_0} < \sum_{i \in \mathscr{I}(\tau_0)} U_i, \tag{2.75}$$

then

$$q\frac{w^*(\tau_0, \tau_1)}{\tau_1 - \tau_0} \geq q\frac{\hat{w}(\tau_0, \tau_1)}{\tau_1 - \tau_0} \geq \sum_{i \in \mathscr{I}(\tau_0)} U_i. \tag{2.76}$$

Hence,

$$g_0 = \max\left\{0, \frac{1}{q} \sum_{i \in \mathscr{I}(\tau_0)} U_i - \frac{w^*(\tau_0, \tau_1)}{\tau_1 - \tau_0}\right\} = 0. \tag{2.77}$$

4. If

$$q\frac{\hat{w}(\tau_0, \tau_1)}{\tau_1 - \tau_0} < \sum_{i \in \mathscr{I}(\tau_0)} U_i \text{ and } \frac{w^*(\tau_0, \tau_1)}{\tau_1 - \tau_0} < \sum_{i \in \mathscr{I}(\tau_0)} U_i, \tag{2.78}$$

then

$$g_0 = \max\left\{0, \frac{\hat{w}(\tau_0, \tau_1)}{\tau_1 - \tau_0} - \frac{w^*(\tau_0, \tau_1)}{\tau_1 - \tau_0}\right\} = 0. \tag{2.79}$$

Hence, $g_0 = 0$ holds when $\hat{w}(\tau_0, \tau_1) < w^*(\tau_0, \tau_1)$. Then $d\Phi/d\tau_0$ remains zero and $\hat{s} \leq qs^*$ by Lemma 2.3. Then, (2.15) always holds by letting $q^2 \leq c$. Therefore, we only consider the case that $\hat{w}(\tau_0, \tau_1) \geq w^*(\tau_0, \tau_1)$ with $q^2 \leq c$. For the speed scaling problem in [15], since there is no constraint of scheduling rate for each individual job, both the online and offline algorithm can always have a solution that only schedule one job that the load intensity gap varies only in at most two time intervals. However, in our problem, since for any PEV, its charging rate can not exceed the maximum charging rate, this leads to that the scheduler should at least charging one PEV at time τ_0. Then we should compute the differential of intensity gap for all intervals and then combine them together. For the time interval $[\tau_0, \tau_1]$, we have

$$\frac{d(\tau_1 - \tau_0)g_0}{d\tau_0}$$

$$= (\tau_1 - \tau_0) \frac{(\tau_1 - \tau_0)\frac{dd(\tau_0, \tau_1)}{d\tau_0} + d(\tau_0, \tau_1)}{(\tau_1 - \tau_0)^2} - g_0^2 \qquad (2.80)$$

$$= \frac{dd(\tau_0, \tau_1)}{d\tau_0}$$

and

$$\frac{d(\tau_1 - \tau_0)g_0^2}{d\tau_0}$$

$$= 2g_0(\tau_1 - \tau_0) \frac{(\tau_1 - \tau_0)\frac{dd(\tau_0, \tau_1)}{d\tau_0} + d(\tau_0, \tau_1)}{(\tau_1 - \tau_0)^2} - g_0^2 \qquad (2.81)$$

$$= 2g_0 \frac{dd(\tau_0, \tau_1)}{d\tau_0} + g_0^2.$$

For the time interval $(\tau_k, \tau_{k+1}], k = 1, 2, \ldots$, we have

$$\frac{d((\tau_{k+1} - \tau_k)g_k)}{d\tau_0} = \frac{dd(\tau_k, \tau_{k+1})}{d\tau_0}$$

$$< \frac{dd(\tau_k, \tau_{k+1})}{d\tau_0}, \qquad (2.82)$$

and

$$\frac{d((\tau_{k+1} - \tau_k)g_k^2)}{d\tau_0} = 2g_k \frac{dd(\tau_k, \tau_{k+1})}{d\tau_0}$$

$$< 2g_0 \frac{dd(\tau_k, \tau_{k+1})}{d\tau_0}, \qquad (2.83)$$

where the last inequality holds because $g_k < g_0, \forall k > 0$. Summing up (2.80), (2.81), (2.82) and (2.83), $d\Phi/d\tau_0$ is upper bounded by

$$\beta_1 a \left(\sum_{k=0}^{\infty} \frac{dd(\tau_k, \tau_{k+1})}{d\tau_0} \right) + \beta_2 b \left(2g_0 \sum_{k=0}^{\infty} \frac{dd(\tau_k, \tau_{k+1})}{d\tau_0} + g_0^2 \right) \tag{2.84}$$

$$= \beta_1 a(-\hat{s} + s^*) + \beta_2 b \left(2g_0(-\hat{s} + s^*) + g_0^2 \right).$$

Then, to prove (2.15), it suffices to show that the following inequality holds, where

$$(a(\hat{s} + l) + b(\hat{s} + l)^2 - (al + bl^2)) + (\beta_2 b(2g_0(-\hat{s} + s^*) + g_0^2) \tag{2.85}$$

$$+ \beta_1 a(-\hat{s} + s^*)) - c(a(s^* + l) + b(s^* + l)^2 - (al + bl^2)) \le 0.$$

It also suffices to show that the following two inequalities hold, where

$$(a + 2bl)\hat{s} + \beta_1 a(-\hat{s} + s^*) - c \cdot (a + 2bl)s^* \le 0, \tag{2.86a}$$

$$b(\hat{s})^2 + \beta_2 b(2g_0(-\hat{s} + s^*) + g_0^2) - c \cdot b(s^*)^2 \le 0. \tag{2.86b}$$

Notice that the LHS of (2.86a) is a linear function of $\hat{s}$. Therefore, it suffices to show that (2.86a) holds for all $s^* \ge 0$ and $g_0 \ge 0$, when $\hat{s} = qg_0$ and $\hat{s} = q(s^* + g_0)$, i.e.,

$$(1 + 2bl/a - \beta_1)qg_0 + (\beta_1 - c(1 + 2bl/a))s^* \le 0 \tag{2.87a}$$

$$(1 - \beta_1)(qg_0 + qs^*) + (\beta_1 - c)s^* \le 0. \tag{2.87b}$$

Since $c \ge 1$ and $q \ge 1$, by setting $\beta_1 = 1 + 2bl/a$, (2.87) holds for all $s^* \ge 0$ and $g_0 \ge 0$. Note that β_1 is finite since $a, b,$ and l are finite number. Similarly, the LHS of (2.86b) is a convex function of $\hat{s}$. Therefore, it suffices to show that (2.86b) holds for all $s^* \ge 0$ and $g_0 \ge 0$ when $\hat{s} = qg_0$ and $\hat{s} = q(s^* + g_0)$. To obtain the lowest competitive ratio, we need to determine the values of q where $1 \le q^2 \le c$ and β_2 that minimize c. This can be achieved by using the numerical method in [14]. We do not present the detailed steps but only the numerical results. That is, the optimal parameters are $q = 1.46$ and $\beta_2 = 2.7$, where the lowest competitive ratio is 2.39. ∎

Proof of Lemma 2.4

We give the proof by contradiction. Note that, in the mth iteration, one of the candidate interval sets is the union of intervals sets $\mathscr{K}^*(m)$ and $\mathscr{K}^*(m+1)$, denoted by $\mathscr{K}'$, i.e.,

$$\mathscr{K}' = \mathscr{K}^*(m) \cup \mathscr{K}^*(m+1). \tag{2.88}$$

Note that the interval set $\mathcal{K}^*(m)$ and $\mathcal{K}^*(m+1)$ have no intersections, i.e.,

$$\mathcal{K}^*(m) \cap \mathcal{K}^*(m+1) = \emptyset, \tag{2.89}$$

then the total energy of $\mathcal{K}'$ is the sum of energy of $\mathcal{K}^*(m)$ and $\mathcal{K}^*(m+1)$, i.e., $y^*(m)\Delta^*(m) + y^*(m+1)\Delta^*(m+1)$. Thus, the total load of the interval set $\mathcal{K}'$ is the balanced the energy over all the intervals in $\mathcal{K}'$, that is,

$$y = \frac{y^*(m)\Delta^*(m) + y^*(m+1)\Delta^*(m+1)}{\Delta^*(m) + \Delta^*(m+1)} \tag{2.90}$$
$$> y^*(m).$$

where the last inequality holds because $y^*(m) < y^*(m+1)$. Thus, it makes a contradiction with $y \leq y^*(m)$ since $y^*(m)$ is the highest total charging rate over all candidate interval sets in iteration m. This completes the proof. ∎

Proof of Theorem 2.2

Proof. For any PEV i, assume that there exist intervals $k_1, k_2, k_3 \in \mathcal{J}(i)$ where $x^*_{ik_1} = 0$, $x^*_{ik_2} \in (0, U_i)$ and $x^*_{ik_3} = U_i$. We separate the proof into the following three parts to match with the three cases of KKT optimality conditions:

1. Interval k_1 must be excluded before interval k_2 and interval k_3 since when schedule $x^*_{ik_1} = 0$ from (2.31), the considered PEV i has not been scheduled that interval k_2 and k_3 should be reserved and goto next iteration. By Lemma 2.4, we have $y^*_{k_1} \geq y^*_{k_2}$ and $y^*_{k_1} \geq y^*_{k_3}$.
2. Interval k_2 must be excluded before interval k_3 since when schedule $x^*_{ik_2}$ and $x^*_{ik_3}$ from (2.29), interval k_2 belongs to the interval set with highest total charging rate and will be excluded in the current iteration, while interval k_3 should be reserved to next iteration. Similarly, by Lemma 2.4 we have $y^*_{k_2} \geq y^*_{k_3}$.
3. For any other interval $k' \in \mathcal{J}(i)$ with $x^*_{ik'} \in (0, U_i)$, $y^*_{k'}$ is the same as $y^*_{k_2}$ because both k' and k_2 belongs to the set $\mathcal{K}^*$ in the same iteration by (2.29) and are assigned the same optimal total charging rate from (2.28).

Therefore, our algorithm satisfies the *KKT* conditions that the solution is always global optimal. ∎

References

1. Z. Ma, D. Callaway, and I. Hiskens, "Decentralized charging control of large populations of plug-in electric vehicles," *IEEE Trans. on Control Systems Technology*, vol. 21, no. 1, pp. 67–78, 2013.
2. K. Mets, T. Verschueren, W. Haerick, C. Develder, and F. D. Turck, "Optimizing smart energy control strategies for plug-in hybrid electric vehicle charging," *in Proc. IEEE/IFIP Netw. Oper. Manage. Symp. (NOMS)*, pp. 293–299, Apr. 2010.
3. Y. He, B. Venkatesh, and L. Guan, "Optimal scheduling for charging and discharging of electric vehicles," *IEEE Trans. on Smart Grid*, vol.3, no.3, pp. 1095–1105, 2012.
4. L. Gan, U. Topcu, and S. H. Low, "Optimal decentralized protocol for electric vehicle charging," *IEEE Trans. on Power System*, vol.28, iss. 2, pp. 940–951, 2012.
5. B. Allan and E. Ran, *Online Computation and Competitive Analysis*, Cambridge, U.K.: Cambridge Univ. Press, 1998.
6. E. Gerding, V. Robu, S. Stein, D. Parkes, A. Rogers, and N. Jennings, "Online mechanism design for electric vehicle charging," *in Proc. of 10th Int. Conf. on Autonomous Agents and Multiagent Systems (AAMAS 2011)*, pp. 811–818, May 2011.
7. M. A. S. Masoum, P. S. Moses, and S. Hajforoosh, "Distribution transformer stress in smart grid with coordinated charging of plug-in electric vehicles," *IEEE Power Energy Syst. Innovative Smart Grid Tech. Conf.*, pp. 1–8, 2012.
8. K. Clement-Nyns , E. Haesen, and J. Driesen, "The impact of charging plug-in hybrid electric vehicles on a residential distribution grid," *IEEE Trans. Power Syst.*, vol. 25, no. 1, pp. 371–380, 2010.
9. S. Chen and L. Tong, "iEMS for large scale charging of electric vehicles architecture and optimal online scheduling," *in Proc. IEEE Int. Conf. Smart Grid Commun. (SmartGridComm)*, pp. 629–634, Nov. 2012.
10. V. Robu, S. Stein, E. H. Gerding, D. C. Parkes, A. Rogers, and N. R. Jennings, "An online mechanism for multi-speed electric vehicle charging," *in Second International Conference on Auctions, Market Mechanisms and Their Applications (AMMA'11)*, pp. 100–112, 2012.
11. S. Stein, E. Gerding, V. Robu, and N. R. Jennings, "A model-based online mechanism with pre-commitment and its application to electric vehicle charging," *in Proce. of the 11th Int. Conf. on Autonomous Agents and Multiagent Systems,* pp. 669–676, 2012.
12. F. Yao, A. Demers, and S. Shenker, "A scheduling model for reduced CPU energy," *in Proc. IEEE Symp. Foundations of Computer Science*, pp. 374–382, 1995.
13. N. Bansal, T. Kimbrel, and K. Pruhs, "Speed scaling to manage energy and temperature," *Journal of the ACM (JACM)*, vol. 54, no. 1, pp. 1–39, 2007.
14. N. Bansal, H. L. Chan, K. Pruhs, and D. Katz, "Improved bounds for speed scaling in devices obeying the cube-root rule," *Proc. 36th Int. Colloqium on Automata, Languages and Programming: Part I*, pp. 144–155, Jul. 2009.
15. N. Bansal, H. L. Chan, and K. Pruhs, "Speed scaling with an arbitrary power function," *In Proc. of the 20th ACM-SIAM Symposium on Discrete Algorithm*, pp. 693–701, 2009.
16. T. W. Lam, L. K. Lee, Isaac K. K. To, and Prudence W. H. Wong, "Speed scaling functions for flow time scheduling based on active job count," *Algorithms-ESA 2008*, pp. 647–659, 2008.
17. S. Boyd and L. Vandenberghe, *Convex Optimization*, Cambridge, U.K.: Cambridge Univ. Press, 2004.
18. Y. Ye, *Interior Point Algorithms: Theory and Analysis*, Wiley-Interscience Press, 1997.
19. A. Ipakchi and F. Albuyeh, "Grid of the future," *IEEE Power and Energy Mag.*, vol. 7, no. 2, pp. 52–62, 2009.
20. A. Santos, A. N. McGuckin, H. Y. Nakamoto, D. Gray, and S. Lis, *Summary of Travel Trends: 2009 National Household Travel Survey*, Federal Highway Administration, Washington, DC, 2011.
21. M. Grant and S. Boyd, CVX: Matlab Software for Disciplined Convex Programming [Online]. Available: http://cvxr.com/cvxMar. 2013, Version 2.0 (beta).

Chapter 3
A MPC-Based PEV Charging Scheduling

Recall that in Chap. 2, the controller of PEV charging station relies on no assumptions nor predictions of the future information. Whereas in this chapter, we study another practical scenario, where the non-causal information about future PEV arrivals is not known in advance, but its statistical information can be estimated. In fact, the statistical information of the future charging demands can often be acquired through historic data, which benefits the control of the PEV charging scheduling in practical scenarios.

3.1 Problem Formulation

In this section, we first introduce the offline PEV charging problem by assuming the knowledge of future information, and then provide the online PEV charging problem. The optimal offline PEV charging scheme will be used as a benchmark to evaluate the performance of the proposed online algorithm.

3.1.1 Optimal Offline PEV Charging Problem

For the offline optimal PEV charging problem, we adopt the same model and notations as Chap. 2 except that (1) the upper bounds of charging rates are assumed to be sufficiently large for all EVs and (2) the charging cost is assumed to be a strictly convex increasing function of the total load. On one hand, the convexity of cost function reflects the fact that each unit of additional power demand becomes more expensive to obtain and make available to the consumer. For example, in the wholesale market, the instantaneous cost can be modeled as a increasing quadratic

© The Author(s) 2017
W. Tang, Y.J.A. Zhang, *Optimal Charging Control of Electric Vehicles in Smart Grids*, SpringerBriefs in Electrical and Computer Engineering,
DOI 10.1007/978-3-319-45862-5_3

function of the instant load [1–3]. On the other hand, it also captures the intent of reducing the load variance over time.

In the ideal case, assume that $l_t, t_i^{(s)}, t_i^{(e)}$, and d_i for all $t = 1, \cdots, T, i \in \mathcal{N}$ are known non-causally at the beginning of the system time. Then, the charging station can solve (3.1) and obtain the optimal charging rate, denoted by x_{it}^* for all time t and the optimal total cost, denoted by Ψ_1. Such a solution is referred to as an "optimal offline solution".

$$
\Psi_1 = \min_{x_{it}} \quad \sum_{t=1}^{T} f\left(\sum_{i \in \mathcal{I}(t)} x_{it} + l_t\right) \tag{3.1a}
$$

$$
\text{s.t.} \quad \sum_{t=t_i^{(s)}}^{t_i^{(e)}} x_{it} = d_i, \forall i \in \mathcal{N}, \tag{3.1b}
$$

$$
x_{it} \geq 0, \forall t = t_i^{(s)}, \cdots, t_i^{(e)}, \forall i \in \mathcal{N}. \tag{3.1c}
$$

In particular, the optimal total charging rate, denoted by s_t^*, is defined as $s_t^* = \sum_{i \in \mathcal{I}(t)} x_{it}^*$.

Note that there are in total $O(T|\mathcal{I}(t)|)$ variables in (3.1), where $|\mathcal{I}(t)|$ denotes the cardinality of the set $\mathcal{I}(t)$. This number can be quite large when the number of cars present at each time slot, $|\mathcal{I}(t)|$, is large. In this subsection, we propose an equivalent transformation of (3.1) that drastically reduces the number of variables. In particular, the following Theorem 3.1 shows that as long as we find the optimal s_t^* $\forall t$, the optimal x_{it}^* $\forall i, t$ can be obtained by earliest deadline first (EDF) scheduling.

Theorem 3.1. *If a set of s_t's satisfy the following inequality for all $n = 1, \cdots, T$*

$$
\sum_{t=1}^{n} \sum_{i \in \{i | t_i^{(e)} = t\}} d_i \leq \sum_{t=1}^{n} s_t \leq \sum_{t=1}^{n} \sum_{i \in \{i | t_i^{(s)} = t\}} d_i, \tag{3.2}
$$

then there exists at least a set of x_{it}'s that is feasible to (3.1). One such set of x_{it}'s can be obtained by EDF scheduling, which charges the PEV $i \in \mathcal{I}(t)$ with the earliest deadline at a rate s_t at each time t. Moreover, when $s_t = s_t^*$, the set of x_{it}'s obtained by EDF scheduling are the optimal solution, x_{it}^*, to (3.1).

Proof. Please see the detailed proof in Appendix "Proof of Theorem 3.1". To see Theorem 3.1, note that (3.2) implies that the total energy charged by any time slot n is no less than the total charging demand that must be satisfied by time n. On the other hand, by EDF scheduling, PEVs with earlier deadlines must be fully charged before those with later deadlines can be charged. Thus, (3.2) guarantees the fulfillment of the charging demands of each individual PEV.

With Theorem 3.1, we can transform (3.1) to the following equivalent problem with T variables.

$$\Psi_1 = \min_{s_t} \quad \sum_{t=1}^{T} f(s_t + l_t) \tag{3.3a}$$

$$\text{s.t.} \quad \sum_{t=1}^{n} s_t \geq \sum_{j=1}^{n} \sum_{i \in \{i | t_i^{(e)} = j\}} d_i, \forall n = 1, \cdots, T. \tag{3.3b}$$

$$\sum_{t=1}^{n} s_t \leq \sum_{j=1}^{n} \sum_{i \in \{i | t_i^{(s)} = j\}} d_i, \forall n = 1, \cdots, T. \tag{3.3c}$$

The optimal solution s_t^* to (3.3) has an interesting feature: it does not change with the cost function $f(s_t + l_t)$, as long as f is strictly convex. Moreover, s_t^* also minimizes the variance of total load subjecting to (3.3b) and (3.3c), where the variance of total load is defined as $\sum_{t=1}^{T}(s_t + l_t - \frac{\sum_{t=1}^{T} s_t + l_t}{T})^2$ [4, 5]. This is proved in Theorem 3.2.

Theorem 3.2. *The optimal solution s_t^* to (3.3) does not change with the cost function $f(.)$, as long as $f(.)$ is strictly convex. Moreover, s_t^* is essentially a load flattening solution that minimizes the variance of total load.*

Proof. Please see the detailed proof in Appendix "Proof of Theorem 3.2".

3.1.2 Online PEV Charging Problem

For the online PEV charging problem, the charging schedule only depends on the statistic information of future load demand, the current based load and the remaining charging demands and deadlines of the PEVs that have arrived so far. In contrast to the offline algorithm that solves (3.3) only once at the beginning of system time, the online charging scheduling algorithm computes the charging rate s_k at each time slot k. The charging rate s_k, once determined, cannot be changed in the future. In particular, s_k is computed by solving a problem similar to (3.3), except that (i) the objective function is now an expectation of charging cost over the random PEV arrivals in the future, (ii) the sum over time in both the objective function and constraints starts from k instead of 1, and (iii) the charging demands at the right hand side of the constraints are replaced by the unfinished charging demands that have not yet been fulfilled by time k. Specifically, the remaining charging demand of PEV i at time k is given by

$$\hat{d}_i^k = d_i - \sum_{t=t_i^{(s)}}^{k-1} x_{it}. \tag{3.4}$$

Note that, $\hat{d}_i^k = d_i$ for all PEVs that have not yet arrived by time $k - 1$. A close look at (3.3) suggests that the charging schedule s_t only depends on the total charging demand that needs to be finished before a certain time, but not the demand due to individual PEVs. Thus, for notational simplicity, we define

$$\tilde{d}_t^k = \sum_{i \in \{i \mid t_i^{(e)} = t\}} \tilde{d}_i^k, \forall t = k, \cdots, T, \tag{3.5}$$

as the total unfinished charging demand at time k that must be completed by time t. With this, we define the state of system at time t as

$$\mathbf{D}_t = [l_t, \tilde{d}_t^t, \tilde{d}_{t+1}^t, \cdots, \tilde{d}_T^t], \tag{3.6}$$

where l_t is the base load at time t, $\tilde{d}_{t'}^t$ is the total unfinished charging demand at time t that must be completed by time t'. Let ξ_t represent the random arrival events at time t. ξ_t is defined as

$$\xi_t = [\iota_t, \eta_t^t, \eta_{t+1}^t, \cdots, \eta_{e_t}^t], \tag{3.7}$$

where ι_t is the base load at time t, $\eta_{t'}^t$ is the total charging demand that arrive at time t and must be fulfilled by time t', e_t is the latest deadline among the PEVs that arrive at time t. Then, the system state at time $t+1$ is defines as

$$\mathbf{D}_{t+1} := g(s_t, \mathbf{D}_t, \xi_{t+1}), \tag{3.8}$$

where $g(.)$ is the transition function between $s_t, \mathbf{D}_t, \xi_{t+1}$ and $\mathbf{D}_{t+1}$. Given $\mathbf{D}_t, s_t$ and ξ_{t+1}, the system state $\mathbf{D}_{t+1}$ can be uniquely determined as follows:

$$l_{t+1} = \iota_{t+1} \tag{3.9}$$

and

$$\tilde{d}_{t'}^{t+1} = \left[\tilde{d}_{t'}^t - \left[s_t - \sum_{j=t}^{t'-1} \tilde{d}_j^t \right]^+ \right]^+ + \eta_{t'}^{t+1}, \forall t' = t+1, \cdots, T. \tag{3.10}$$

Here, $[x]^+ = \max\{x, 0\}$. With the above definitions of system state and state transition, we are now ready to rewrite (3.3) into its online counterpart. In particular, given D_k at a current time slot k, the optimal online charging decision s_k is the solution to the following finite-horizon dynamic programming problem.

$$Q_k(\mathbf{D}_k) = \min_{s_k} \quad f(s_k + l_k) + \mathbf{E}_{\xi_{k+1}}[Q_{k+1}(g(s_k, \mathbf{D}_k, \xi_{k+1}))] \tag{3.11a}$$

$$\text{s. t.} \quad \tilde{d}_k^k \leq s_k \leq \sum_{t=k}^T \tilde{d}_t^k, \tag{3.11b}$$

where $Q_{k+1}(g(s_k, \mathbf{D}_k, \xi_{k+1}))$ is the optimal value of the dynamic programming at time $k + 1$. The left side of (3.11b) ensures all changing demands to be satisfied before their deadlines. The right side of (3.11b) implies that the total charging power up to a certain time cannot exceed the total demands that have arrived up to that time. By slight abuse of notation, in the rest of the chapter we denote the optimal solutions to both the online and offline problems as s_k^*, when no confusion arises. The actual meaning of s_k^* will be clear from the context. Suppose that s_k^* is the optimal solution to (3.11) at stage k. Then, the resultant total charging cost, denoted by Ψ_2, is

$$\Psi_2 = \sum_{k=1}^{T} f(s_k^* + l_k). \tag{3.12}$$

Note that (3.11a) comprises nested expectations with respect to random PEV arrivals at each time slot. Except for few special cases, it is hard to provide the closed-form of the optimal solution to (3.11). Then, (3.11) can be solved by the commonly numerical methods, such as backward reduction and the sample average approximation (SAA) based on Monte Carlo sampling techniques [6–9]. Notice that (3.11) includes continuous spaces of both state and variable, where the continuous spaces should be discretized into indefinitely small pieces in the commonly numerical methods. Thus, the discretization of continuous spaces and the curse of dimensionality lead to a prohibitive complexity of the commonly numerical methods.

3.2   Related Work

The works on the PEV charging scheduling with uncertain PEV load demand include both simulation-based evaluations [10, 11] and theoretical performance guarantees [4, 5, 12]. Meanwhile, MPC is one of most commonly approaches for which has been widely adopted in recent studies [4, 5, 11, 12]. Rao and Yao [11] leverages the MPC based method to design a dynamic charging and driving cost control scheme. Both [4] and [5] apply MPC algorithms to minimize the load variation. Bansal et al. [12] proposes a plug and play MPC approach to minimize the voltage fluctuations by assuming that the load demand is time-periodic. Compared to [4, 5, 11, 12], in this chapter we analyze the performance gap between the solution of MPC approach and the optimal solution regardless of the distribution of the load demand. Besides, we provide a more scalable algorithm with $O(1)$-complexity as well as the optimality analysis for the case when the load demand is first-order periodic. Additionally, the objective functions in [4, 5, 11, 12] are quadratic forms of load demand. Whereas in this chapter, the objective function is a general strictly convex increasing function which reflects both the charging cost and the load variance.

In this chapter, we provide a Model Predictive Control (MPC) based algorithm to solve the online charging scheduling problem. In contrast to the previous work, we rigorously prove that the proposed algorithm yields a near-optimal solution that has a bounded performance gap from the optimal solution regardless of the distribution of exogenous random variables. Furthermore, our rigorous analysis shows that the proposed algorithm can be made scalable when the random process describing the arrival of charging demands is first-order periodic. That is, the complexity of proposed algorithm can be reduced to $O(1)$, which is independent of T. Extensive simulations show that the proposed online algorithm performs very closely to the optimal online algorithm. The performance gap is smaller than 0.4% in most cases. As such, the proposed online algorithm is very appealing for practical implementation due to its scalable computational complexity and close to optimal performance.

3.3   MPC-Based Online Charging Algorithm

Instead of adopting the commonly numerical methods to solve (3.11), we are motivated to solve a much simpler problem here: the one obtained by replacing all exogenous random variables by their expected values. This is referred to the expected value problem [7–9] and the MPC approach [4, 5, 11, 12]. Moreover, by exploring the load flattening feature of the optimal solution to the expected value problem, we propose a low-complexity online Expected Load Flattening (ELF) algorithm, as shown in Sect. 3.3.1. Section 3.3.2 proves that the optimal solution to the expected value problem yields a bounded performance gap compared with the optimal solution to (3.11). Numerical results show that the performance gap is negligible ($< 0.4\%$) in most cases.

3.3.1   Algorithm Description

Denote the expectation of ξ_t as

$$\mu_t = [v_t, \mu_t^t, \cdots, \mu_T^t], \tag{3.13}$$

where

$$v_t = \mathbf{E}[\iota_t], \mu_{t'}^t = \mathbf{E}[\eta_{t'}^t], \forall t' = t, \cdots, T. \tag{3.14}$$

Replacing ξ_t in (3.11) with μ_t, we obtain the following deterministic problem:

$$\min_{s_k} \quad f(s_k + l_k) + \sum_{t=k+1}^{T} f(s_t + v_t) \tag{3.15a}$$

$$\text{s. t.} \quad \sum_{t=k}^{j} s_t \geq \sum_{t=k}^{j} \tilde{d}_t^k + \sum_{m=k+1}^{j} \sum_{n=m}^{j} \mu_n^m, \forall j = k, \cdots, T, \tag{3.15b}$$

$$\sum_{t=k}^{j} s_t \leq \sum_{t=k}^{T} \tilde{d}_t^k + \sum_{m=k+1}^{j} \sum_{n=m}^{e_m} \mu_n^m, \forall j = k, \cdots, T, \tag{3.15c}$$

In each time k, we solve problem (3.15) and obtain the optimal charging solution s_k^*. Then, problem (3.15) is resolved with the updated $\tilde{d}_t^k$ according to the realization of the PEVs arrived in next time. So on and so forth, we obtain the optimal charging solution s_k^* for time stage $k = 2, \cdots, T$. The total cost, denoted by Ψ_3, is defined as

$$\Psi_3 = \sum_{k=1}^{T} f(s_k^* + l_k), \tag{3.16}$$

where s_k^* is the optimal solution to (3.15) at time stage k. The solution to (3.15) is always feasible to (3.11) in the sense that it always guarantees fulfilling the charging demand of the current parking PEVs before their departures. This is because the constraints of s_k in (3.11) are included in (3.15).

Due to the convexity of $f(\cdot)$, the optimal solution is the one that flattens the total load as much as possible. By exploiting the load flattening feature of the solution, we present in Algorithm 2 the online ELF algorithm that solves (3.15) with complexity $O(T^3)$. The online ELF algorithm have a lower computational complexity than generic convex optimization algorithms, such as the interior point method, which has a complexity $O(T^{3.5})$ [13]. Notice that similar algorithms have been proposed in the literature of speed scaling problems [14, 15] and PEV charging problems [16]. The optimality and the complexity of the algorithm have been proved therein, and hence omitted here. The algorithm presented here, however, paves the way for further complexity reduction to $O(1)$ in Sect. 3.4. For notation brevity, we denote in the online ELF algorithm

$$\bar{d}_{t''}^{t'} = \begin{cases} \tilde{d}_{t''}^{t'}, & \text{for } t'' = k, \cdots, T, t' = k, \\ \mu_{t''}^{t'}, & \text{for } t'' = t', \cdots, T, t' = k+1, \cdots, T. \end{cases} \tag{3.17}$$

The key idea of online algorithm ELF is to balance the charging load among all time slots $k, \cdots, T$ and then assign the balanced load at time k to the solution s_k^*. Specifically, step 3–5 is to find the time interval that has the maximum load density during time k to T, and set the optimal charging rate for that time interval to be equal to the maximum density. The time interval is then deleted, and the process is repeated until the current time k belongs to the maximum-density interval.

Algorithm 2: Online algorithm ELF
---

    input : $\mathbf{D}_k, \mu_t, t = k+1, \cdots, T$

    output: s_k

1 initialization $i = 0, j = 0$;

2 **repeat**

3   For all time slot $i = k, \cdots, T, j = i, \cdots, T$, compute

$$i^*, j^* = \arg\max_{i,j}\left\{ \frac{\sum_{t'=i}^{j}(\sum_{t''=t'}^{j} \bar{d}_{t''}^{t'} + v_{t'})}{j - i + 1} \right\}. \tag{3.18}$$

4   Set

$$y^* = \frac{\sum_{t'=i^*}^{j^*}(\sum_{t''=t'}^{j^*} \bar{d}_{t''}^{t'} + v_{t'})}{j^* - i^* + 1}. \tag{3.19}$$

5   Delete time slot $i^*, \cdots, j^*$ and relabel the existing time slot $t > j^*$ as $t - j^* + i^* - 1$.

6 **until** $i^* = k$;

7 Set $s_k = y^* - l_k$.

3.3.2   Optimality Analysis

In this subsection, we analyze the optimality of the solution to (3.15). A well-accepted metric, *Value of the Stochastic Solution* (VSS) is adopted to evaluate optimality gap between the optimal online solution and the solution to the expected value problem [7–9]. To evaluate the VSS, the previous study, e.g., [8, 9], mainly focus on the numerical simulations. Whereas in this subsection, we derive an upper bound of the VSS through the rigorous theoretical analysis. we denoted by Ξ a scenario, which is defined as a possible realization of the sequence of random load demand [6],

$$\Xi = [\xi_2, \xi_3, \cdots, \xi_T]. \tag{3.20}$$

Here, we treat ξ_1 as deterministic information since the demand of PEVs arrived at the first stage is known by the scheduler. Let Φ_1, Φ_2 and Φ_3 be the expectation of the optimal value of the offline problem (3.3), the online problem (3.11) and the expected value problem (3.15) over the set of scenario Ξ, respectively. That is,

$$\Phi_1 = E_\Xi[\Psi_1(\Xi)], \Phi_2 = E_\Xi[\Psi_2(\Xi)], \Phi_3 = E_\Xi[\Psi_3(\Xi)]. \tag{3.21}$$

It has been proved previously [7, 8] that

$$\Phi_1 \leq \Phi_2 \leq \Phi_3. \tag{3.22}$$

To assess the benefit of knowing and using the distributions of the future outcomes, the VSS is defined as

$$\text{VSS} = \Phi_3 - \Phi_2. \tag{3.23}$$

To show that the online algorithm ELF yields a bounded VSS, we need to bound Φ_3 and Φ_2. Generally, it is hard to calculate Φ_2 or analyze the lower bound of Φ_2 directly [8, 9]. Thus, we choose to analyze the lower bound of Φ_1 instead, since (3.22) shows that the lower bound of Φ_1 is also the bound of Φ_2. In other words, for the gap VSS, we have

$$\text{VSS} = \Phi_3 - \Phi_2 \leq \Phi_3 - \Phi_1. \tag{3.24}$$

In what follows, we will derive in Proposition 3.1 the lower bound of Φ_1. Likewise, we will also derive the upper bound of Φ_3 in Proposition 3.2.

Proposition 3.1.

$$\Phi_1 \geq Tf\left(\frac{\sum_{t=1}^{e_1} \tilde{d}_t^1 + \sum_{t=2}^{T} \sum_{j=t}^{e_t} \mu_t^j + \sum_{t=1}^{T} v_t}{T}\right). \tag{3.25}$$

Proof. Please see the detailed proof in Appendix "Proof of Proposition 3.1".

Let $\mathcal{O}(t)$ be the set that

$$\mathcal{O}(t) = \{(m,n)|e_m \geq t, m = 1, \cdots, t, n = t, \cdots, e_m\}. \tag{3.26}$$

Then we show that Φ_3 is bounded by the following proposition.

Proposition 3.2. *For any distribution of $\xi_t, t = 1, \cdots, T$, there is*

$$\Phi_3 \leq \text{E}\left[\sum_{t=1}^{T} f\left(\sum_{(m,n) \in \mathcal{O}(t)} \eta_n^m + \iota_t\right)\right]. \tag{3.27}$$

Proof. Please see the detailed proof in Appendix "Proof of Proposition 3.2".

Now, we are ready to present Theorem 3.3, which states that the VSS is bounded for any distribution of random variables.

Theorem 3.3. *For any distribution of random vector ξ_t, $t = 1, \cdots, T, n = t, \cdots, T$, there is*

$$\text{VSS} \leq \text{E}\left[\sum_{t=1}^{T} f\left(\sum_{(m,n) \in \mathcal{O}(t)} \eta_n^m + \iota_t\right)\right] - Tf\left(\frac{\Gamma}{T}\right), \tag{3.28}$$

where $\Gamma = \sum_{t=1}^{e_1} \tilde{d}_t^1 + \sum_{t=2}^{T} \sum_{j=t}^{e_t} \mu_t^j + \sum_{t=1}^{T} v_t$.

Proof. Please see the detailed proof in Appendix "Proof of Theorem 3.3".

Before leaving this section, we would like to comment that in practice, the performance gap between the online algorithm ELF and the optimal online algorithm is often much smaller than the bound of VSS. This will be elaborated in the numerical results in Sect. 3.5.

3.4   Online Algorithm ELF Under First-Order Periodic Process

Notice that the complexity of $O(T^3)$ of online algorithm ELF mainly comes from step 3, which exhaustively searches the maximum-density period $[i^*,j^*]$ over all subintervals within $[k,T]$. When the arrival process is first-order periodic, we argue that the searching scope can be reduced to one period from the whole system time T. Thus, the complexity of step 3 is limited by the length of a period instead of T. As a result, the complexity of the algorithm reduces from $O(T^3)$ to $O(1)$, implying that it does not increase with the system time T, and thus the algorithm is perfectly scalable. In practice, the arrival process of the charging demands are usually periodic. For example, the arrival of charging demands at a particular location is statistically identical at the same time every day during weekdays (or weekends). In Sect. 3.4.1, we will investigate a special case when the random arrival process is first-order stationary. The investigation is then generalized to the first-order periodic case in Sect. 3.4.2.

3.4.1   First-Order Stationary Process

In this subsection, we show that the optimal solution to (3.15) can be calculated in closed form if the arrival process is first-order stationary. By first-order stationary, we mean that the statistical mean of ξ_t, i.e., v_t and $\mu_t^{t'}, t' = t, \cdots, T$ only depends on the relative time difference $\tau = t' - t$, but not the absolute value of t. We can then replace v_t by v and replace $\mu_{t'}^t$ by μ_τ, where $\tau = t' - t$. When the arrival process is first-order stationary, i.e., $\forall t = 2, \cdots, T$, μ_t in (3.13) is no longer a function of t, and can be represented as

$$\mu = [v, \mu_1, \mu_2, \cdots, \mu_{\bar{e}}, 0, \cdots, 0], \tag{3.29}$$

where $\bar{e}$ is the maximum parking time of a PEV. To find the subinterval $[i^*,j^*] \subseteq [k,T]$ with the maximum density, we decompose the search region $\{i,j | i = k, \cdots, T, j = i, \cdots, T\}$ into three sub-regions, i.e., $\{i,j | i = k, j = k, \cdots, k + \bar{e}\}$,

$\{i,j|i=k,j=k+\bar{e}+1,\cdots,T\}$ and $\{i,j|i=k+1,j=i,\cdots,T\}$. Let $[i_1,j_1]$ be the maximum-density subinterval within the first sub-region $\{i,j|i=k,j=k,\cdots,k+\bar{e}\}$, and the corresponding maximum density is denoted as X. Likewise, let $[i_2,j_2]$ be the maximum-density subinterval of the second sub-region $\{i,j|i=k,j=k+\bar{e}+1,\cdots,T\}$, with the maximum density denoted as Y, and $[i_3,j_3]$ be the maximum-density subinterval within the third sub-region $\{i,j|i=k+1,j=i,\cdots,T\}$, with the maximum density denoted as Z. By definition, $i_1=i_2=k$. The maximum density X of the first sub-region can be obtained by searching j_1 over $\{k,\cdots,k+\bar{e}\}$:

$$X = \max_{k\le n\le k+\bar{e}} \left\{ \frac{\sum_{t=k}^n \tilde{d}_t^k + \sum_{j=1}^n (n-k-j+1)\mu_j + l_k - v}{n-k+1} + v \right\}. \tag{3.30}$$

The searching complexity is limited by $\bar{e}$ instead of the system time T. Moreover, we will show in the following Lemma 3.1 that Y and Z can be calculated in closed form. That is, the complexity of obtaining the maximum densities over the second and third sub-regions is very low.

Lemma 3.1. *The maximum density of* $\{i,j|i=k,j=k+\bar{e}+1,\cdots,T\}$ *is achieved by setting* $i_2=k,j_2=T$, *and calculated by*

$$Y = \frac{\sum_{t=k}^{k+\bar{e}} \tilde{d}_t^k + \sum_{j=1}^{k+\bar{e}} (T-k-j+1)\mu_j + l_k - v}{T-k+1} + v. \tag{3.31}$$

Moreover, the maximum density of $\{i,j|i=k+1,j=i,\cdots,T\}$ *is achieved by setting* $i_3=k+1,j_3=T$, *and calculated by*

$$Z = \frac{\sum_{j=1}^{k+\bar{e}} (T-k-j+1)\mu_j}{T-k} + v. \tag{3.32}$$

Proof. Please see the detailed proof in Appendix "Proof of Lemma 3.1".

The largest of X, Y, and Z is the maximum density of the interval $[i^*,j^*] \subseteq [k,T]$ over all possible pairs $i,j \in \{i=k,\cdots,T,j=i,\cdots,T\}$. Specially, if X or Y is the largest one, then k is already contained in the maximum-density interval, and thus X or Y is the optimal charging rate at time k. On the other hand, if Z is the largest, then the maximum-density interval, i.e., $[k+1,T]$, does not include k. Following Algorithm 2, we will delete the maximum-density interval and repeat the process. Now, time slot k is the only remaining time slot after deletion. This implies that all charging demands that have arrived by time slot k should be fulfilled during time slot k. These arguments are summarized in Proposition 3.3, which provides the closed form solution to (3.15).

Proposition 3.3. *The optimal charging schedule to (3.15) is given by the following close-form*

$$s_k^* = \begin{cases} X - l_k, & \text{if } X = \max\{X,Y,Z\}, & (3.33) \\ Y - l_k, & \text{if } Y = \max\{X,Y,Z\}, & (3.34) \\ \sum_{t=k}^{k+\bar{e}} \tilde{d}_t^k, & \text{otherwise.} & (3.35) \end{cases}$$

It is obvious that the complexity of calculating s_k^* is independent of T. In other words, the computational complexity is reduced to $O(1)$.

3.4.2 First-Order Periodic Process

In this subsection, we extend Proposition 3.3 to the case when the arrival process is first-order periodic. By first-order periodic, we mean that μ_t in (3.13) repeats itself periodically. Suppose that the period is p. Then, instead of considering μ_t for $t = k+1, \cdots, T$, we only need to consider μ_t for one period, i.e., for $t = k+1, k+p$:

$$\mu_{k+1} = [\nu_{k+1}, \mu_{k+1}^{k+1}, \mu_{k+2}^{k+1}, \cdots, \mu_{k+e_1}^{k+1}, 0, \cdots, 0],$$

$$\vdots \tag{3.36}$$

$$\mu_{k+p} = [\nu_{k+p}, \mu_{k+p}^{k+p}, \mu_{k+p+1}^{k+p}, \cdots, \mu_{k+e_p}^{k+p}, 0, \cdots, 0].$$

Here, $e_n \leq T, n = 1, \cdots, p$ is the maximum parking time for PEVs arriving at time $k + n$. Specially, we define $\hat{e}$ as $\hat{e} = \max\{e_{k+1}, e_{k+2}, \cdots, e_{k+p}\}$. Similar to the first-order stationary case, we decompose the search region $\{i,j | i = k, \cdots, T, j = i, \cdots, T\}$ into three sub-regions, i.e., $\{i,j | i = k, j = k, \cdots, k+\hat{e}\}$, $\{i,j | i = k, j = k+\hat{e}+1, \cdots, T\}$ and $\{i,j | i = k+1, j = i, \cdots, T\}$. Let $[\hat{i}_1, \hat{j}_1]$ be the subinterval with the maximum density, denoted by $\hat{X}$, over the first sub-region $\{i,j | i = k, j = k, \cdots, k+\hat{e}\}$, $[\hat{i}_2, \hat{j}_2]$ be the subinterval with the maximum density, denoted by $\hat{Y}$, over the second sub-region $\{i,j | i = k, j = k+\bar{e}+1, \cdots, T\}$ and $[\hat{i}_3, \hat{j}_3]$ be the subinterval with the maximum density, denoted by $\hat{Z}$, over the third sub-region $\{i,j | i = k+1, j = i, \cdots, T\}$. By definition, $\hat{i}_1 = \hat{i}_2 = k$. Similar to the stationary case, $\hat{X}$ can be calculated by searching $\hat{j}_1$ over $\{k, \cdots, k+\hat{e}\}$. That is,

$$\hat{X} = \max_{k \leq t \leq k+\hat{e}} \frac{\sum_{n=k}^{t}(\tilde{d}_n^k + \nu_n) + \sum_{n=k}^{t}\sum_{m=n}^{t}\mu_m^n}{n - k + 1}. \tag{3.37}$$

Moreover, Lemma 3.2 shows that $\hat{Y}$ and $\hat{Z}$ can be calculated once the maximum density of $[k+1, k+\hat{e}]$ has been obtained. Here, we define $[\bar{i}, \bar{j}]$ to be the maximum-density interval within $[k+1, k+\hat{e}]$, i.e.,

$$\bar{i},\bar{j} = \operatorname*{arg\,max}_{k+1 \le i \le j \le k+\hat{e}} \frac{\sum_{n=i}^{j}(\sum_{m=n}^{k+e_n}\mu_m^n + v_n)}{j-i+1}. \tag{3.38}$$

Now we are ready to present Lemma 3.2.

Lemma 3.2. *The maximum density of* $\{i,j|i=k,j=k+\hat{e}+1,\cdots,T\}$ *is calculated by*

$$\hat{Y} = \frac{\sum_{n=k}^{\hat{j}_2}(\sum_{m=n}^{k+e_n}\mu_m^n + v_n)}{\hat{j}_2 - k + 1}, \tag{3.39}$$

where

$$\hat{j}_2 = \begin{cases} \max\{\bar{j}, k+\hat{e}+1\}, & \text{if } \bar{j} < \bar{i}+p, \\ \bar{j} + (r-1)p, & \text{otherwise.} \end{cases} \tag{3.40}$$

The maximum density of $\{i,j|i=k+1,j=i,\cdots,T\}$ *is calculated by*

$$\hat{Z} = \frac{\sum_{n=\hat{i}_3}^{\hat{j}_3}(\sum_{m=n}^{k+e_n}\mu_m^n + v_n)}{\hat{j}_3 - \hat{i}_3 + 1}, \tag{3.41}$$

where $\hat{i}_3 = \bar{i}$ *and*

$$\hat{j}_3 = \begin{cases} \bar{j}, & \text{if } \bar{j} < \bar{i}+p, \\ \bar{j} + (r-1)p, & \text{otherwise.} \end{cases} \tag{3.42}$$

Proof. Please see the detailed proof in Appendix "Proof of Lemma 3.2".

Based on Lemma 3.2, we can modified the searching region of step 3 of online algorithm ELF as follows:

- if $\bar{j} < \bar{i}+p$, the interval with the maximum density during time stages $[k+1,T]$ is $[\bar{i},\bar{j}]$. Then, for the step 3 of online algorithm ELF, the search region of i,j is reduced from $\{i,j|i=k,\cdots,T,j=i,\cdots,T\}$ to $\{i,j|i=k,\cdots,\bar{i},j=i,\cdots,\bar{i},\bar{j}\}$, where $\bar{i} \in [k+1,k+\hat{e}]$.
- If $\bar{j} \ge \bar{i}+p$, the interval with the maximum density during time stages $[k+1,T]$ is $[\bar{i},\bar{j}+(r-1)p]$. Then, for the step 3 of online algorithm ELF, the search region of i,j can be reduced from $\{i,j|i=k,\cdots,T,j=i,\cdots,T\}$ to $\{i,j|i=k,\cdots,\bar{i},j=i,\cdots,\bar{i},\bar{j}+(r-1)p\}$, where $\bar{i} \in [k+1,k+\hat{e}]$.

Hence, the searching region of the modified online algorithm ELF is only related to $[k+1,k+\hat{e}]$ instead of T. Thus, the computational complexity of the online algorithm ELF is $O(1)$ instead of $O(T^3)$ when the arrival process is first-order periodic.

3.5 Simulations

In this section, we investigate the performance of the proposed online algorithm ELF through numerical simulations. All the computations are solved in MATLAB on a computer with an Intel Core i3-2120 3.30 GHz CPU and 8 GB of memory. For comparison purpose, we also simulate the performance of the optimal offline solution, the optimal online solution obtained by SAA method as well as the online algorithm AVG, which charges each PEV as the ratio of its charging demand to parking time [3]. The average cost of online algorithm AVG is denoted by Φ_4. Define the relative performance loss of the online algorithm ELF and AVG compared with the optimal online solution as $\frac{\Phi_3 - \Phi_2}{\Phi_2}$ and $\frac{\Phi_4 - \Phi_2}{\Phi_2}$, respectively. Similar to [3, 16], we adopt a quadratic cost function in the simulations, i.e., $f(s_t + l_t) = (s_t + l_t)^2$.

3.5.1 Average Performance Evaluation

In this subsection, we evaluate the average performance of the online algorithm ELF under three different traffic patterns, i.e., light, moderate, and heavy traffics. In particular, the system time T is set to be 24 h. The PEV arrivals follow a Poisson distribution and the parking time of each PEV follows an exponential distribution [16]. The mean arrival and parking durations of the three traffic patterns are listed in Table 3.1. The main difference lies in the arrival rates at the two peak hours, i.e. 12:00 to 14:00 and 18:00 to 20:00. The settings of the peak hour match with the realistic vehicle trips in National Household Travel Survey (NHTS) 2009 [17]. Specially, the average number of total PEVs simulated in scenario 1, 2 and 3 are 104, 204 and 304, respectively. We choose the base load profile of one day in the service area of South California Edison from [18]. Each PEV's charging demand is uniformly chosen from [25, 35] kWh.

For each scenario, we simulate the average performance of 10^5 independent instances by adopting optimal offline algorithm, optimal online algorithm, online algorithm ELF and AVG, respectively. The average total load over time are plotted

Table 3.1 Parameter settings of the PEV traffic patterns

Time of day	Arrival rate (PEVs/hour)			Mean parking Time (hour)
	S. 1	S. 2	S. 3	
08:00–10:00	7	7	7	10
10:00–12:00	5	5	5	1/2
12:00–14:00	10	35	60	2
14:00–18:00	5	5	5	1/2
18:00–20:00	10	35	60	2
20:00–24:00	5	5	5	10
24:00–08:00	0	0	0	0

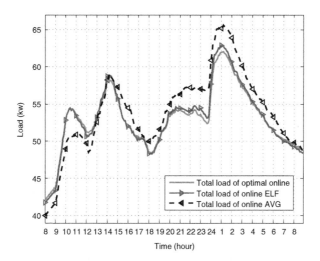

Fig. 3.1 Base load and total load of four algorithms in Scenario 1

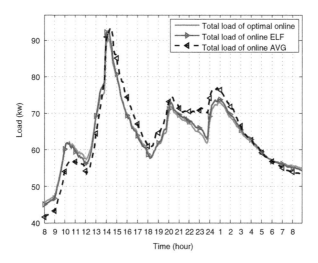

Fig. 3.2 Base load and total load of four algorithms in Scenario 2

in Figs. 3.1, 3.2 and 3.3. For each scenario, we calculate the average costs of four algorithms normalized by the optimal offline, respectively, and plot the normalized costs in Fig. 3.4. In addition, the VSS and the relative performance loss are shown in Table 3.2.

From Figs. 3.1, 3.2 and 3.3, we notice that the curve of total load output by the online algorithm ELF follows closely to that of optimal offline algorithm. Figure 3.4 and Table 3.2 show that the online algorithm ELF has on average less than 7 % extra cost compared with the optimal offline algorithm throughout three scenarios. Moreover, the online algorithm ELF performs very closed to the optimal online

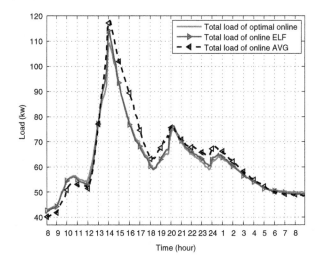

Fig. 3.3 Base load and total load of four algorithms in Scenario 3

Table 3.2 Average
performance comparison
under three traffic patterns

Scenario	VSS	$\frac{\Phi_3 - \Phi_2}{\Phi_2}$	$\frac{\Phi_4 - \Phi_2}{\Phi_2}$
1	0.1178	0.19 %	3.50 %
2	0.1319	0.28 %	4.46 %
3	0.1536	0.38 %	5.82 %

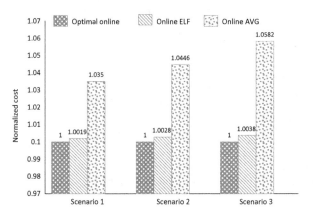

Fig. 3.4 The average costs of four algorithms normalized by the average cost of the optimal offline
in three scenarios

algorithm. the VSS and the relative performance loss are no more than 0.1536
and 0.38 % respectively. In contrast, online algorithm AVG largely deviate from
the curve of the optimal offline algorithm in all the three scenarios. Compared
with online algorithm AVG, online algorithm ELF always performs better in three
scenarios, which produces half of the extra cost compared with the optimal offline

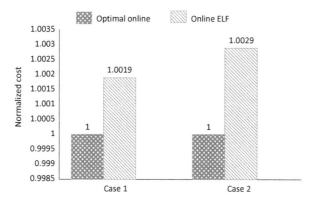

Fig. 3.5 The average costs of three algorithms normalized by the average cost of the optimal offline in two cases

algorithm than that produced by AVG. This is because AVG does not rely on any future information that leads to a larger fluctuation of the total load curve.

3.5.2 Discussion the Influence of Variance

Note that only the statistic means of the random variables are required in the online algorithm ELF. Intuitively, the variance of the random variables may influence the results. In this subsection, we discuss how the variance of charging demands affects the performance of our online algorithm. Let the arrivals and parking durations of PEVs be the same as scenario 1 in Table 3.1. For comparison, we simulate two cases where each PEV's charging demand is uniformly chosen from different intervals: case 1: $[25, 35 \text{ kWh}]$, where the variance is $\frac{25}{3}$ (kWh^2); case 2: $[5, 35 \text{ kWh}]$, where the variance is $\frac{225}{3}$ (kWh^2). Thus, the variance of the charging demands in case 2 is 9 times of that in case 1. For each scenario, we simulate the average performance of 10^5 independent instances by adopting four algorithms respectively. For each scenario, we calculate the average costs of four algorithms normalized by the optimal offline, respectively, and plot the normalized costs in Fig. 3.5. Figure 3.5 shows that the online algorithm ELF has on average less than 3.6 % extra cost compared with the optimal offline algorithm and 0.29 % extra cost compared with the optimal online algorithm. Hence, when the variance of random variables increase 8 times, the performance gap between the online algorithm ELF and optimal offline algorithm changes at most 0.5 % and the performance gap between the online algorithm ELF and optimal online algorithm changes at most 0.1 %.

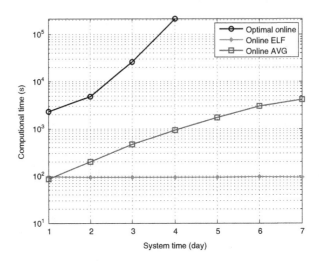

Fig. 3.6 CPU computational time over system time for three online algorithms

3.5.3   Complexity Comparison

In this subsection, we proceed to evaluate the computational complexity of the online ELF algorithm when the load demand is first order periodic. Similar to Sect. 3.4.2, we use the optimal online algorithm and the online AVG algorithm as performance benchmarks. One period is set to be one day. For each day, the settings of traffic patterns are adopted the same as scenario 1 shown in Table 3.1, and other parameter settings are adopted the same as Sect. 3.5.1. We simulate 7 cases in total, where the system time are set to be $1, 2, \cdots, 7$ days and average number of PEVs per day is 204. For each algorithm, we record the CPU computational time at each time stage and calculate the average CPU computational time as the sum of CPU computational times of all time stages divided by the number of time stages. Each point in Fig. 3.6 is an average of 100 independent instances. Figure 3.6 shows that the CPU computational time of the online ELF algorithm is almost a constant around 100 s regardless of the total system time. This observation is consistent with our analysis in Sect. 3.4.2, which states that the algorithm is scalable. In comparison, the average CPU computational time of the optimal online algorithm and online AVG algorithm grows very quickly as system time increases. We notice that the optimal online algorithm consumes more than 2.5 days when system time increases to 4 days, while online AVG algorithm takes around 15.5 min. It is foreseeable that the computational complexity of optimal online algorithm will become extremely expensive as we further increase the system time.

3.6 Conclusions

In this chapter, we formulate the optimal PEV charging scheduling problem as a finite-horizon dynamic programming problem. Instead of adopting the backward reduction method with prohibitive complexity, we provide a MPC-based online algorithm with a computational complexity of $O(T^3)$, where T is the total number of the time stages. We rigorously show that the performance gap between the proposed online algorithm and the optimal solution is bounded regardless of the distribution of random variables. Moreover, we show that the algorithm can be made scalable with $O(1)$-complexity when the arrivals of load demands are first-order periodic. Our analyses are validated through extensive simulations, which shows the proposed online algorithm can efficiently minimize the charging cost and reduce the load variance.

Appendix

Proof of Theorem 3.1

We use the inductive method to show that through EDF scheduling, all the PEVs can be fulfilled charging before deadlines. For $n = 1$, (3.2) becomes

$$\sum_{i \in \{i | t_i^{(s)} = 1\}} d_i \geq s_1 \geq \sum_{i \in \{i | t_i^{(e)} = 1\}} d_i. \tag{3.43}$$

Thus, by EDF scheduling, we can first satisfy the demand of PEVs whose deadline at time stage 1. That is, for any PEV $i \in \{i | t_i^{(e)} = 1\}$, we set

$$x_{i1} = d_i. \tag{3.44}$$

Assuming that for all time stage m, EDF scheduling can fulfill charge all the PEVs which depart at or before time stage m, i.e., there exists at least a set of x_{it}'s that satisfy

$$\sum_{t=t_i^{(s)}}^{t_i^{(e)}} x_{it} = d_i, \forall i \in \{i | t_i^{(e)} \leq m\}, \tag{3.45a}$$

$$x_{it} \geq 0, \forall t = t_i^{(s)}, \cdots, t_i^{(e)}, \forall i \in \{i | t_i^{(e)} \leq m\}. \tag{3.45b}$$

Since

$$\sum_{t=1}^{m} s_t \geq \sum_{t=1}^{m} \sum_{i \in \{i | t_i^{(e)} = t\}} d_i, \tag{3.46}$$

then, $\sum_{t=1}^{m} s_t - \sum_{t=1}^{m} \sum_{i \in \{i | t_i^{(e)} = t\}} d_i$ represents the amount of power which is outputted from the charging station during time stage $1, \cdots, m$ and charged to the PEVs with deadline after time stage m. By EDF scheduling, once the PEVs which depart at time m have been fulfilled charging, we will first charge the PEVs which depart at time stage $m + 1$. Thus, if

$$\sum_{t=1}^{m} s_t - \sum_{t=1}^{m} \sum_{i \in \{i | t_i^{(e)} = t\}} d_i \geq \sum_{i \in \{i | t_i^{(e)} = m+1\}} d_i, \tag{3.47}$$

we finish charging of PEVs with deadline $m + 1$, and then go to charge the PEVs with deadline $m + 2$. If

$$\sum_{t=1}^{m} s_t - \sum_{t=1}^{m} \sum_{i \in \{i | t_i^{(e)} = t\}} d_i < \sum_{i \in \{i | t_i^{(e)} = m+1\}} d_i, \tag{3.48}$$

then the PEVs with deadline $m + 1$ have been charged as power $\sum_{t=1}^{m} s_t - \sum_{t=1}^{m} \sum_{i \in \{i | t_i^{(e)} = t\}} d_i$. At time stage $m + 1$. Since

$$\sum_{t=1}^{m+1} s_t \geq \sum_{t=1}^{m+1} \sum_{i \in \{i | t_i^{(e)} = t\}} d_i, \tag{3.49}$$

then,

$$s_{m+1} \geq \sum_{i \in \{i | t_i^{(e)} = m+1\}} d_i - \left(\sum_{t=1}^{m} s_t - \sum_{t=1}^{m} \sum_{i \in \{i | t_i^{(e)} = t\}} d_i \right), \tag{3.50}$$

which means all the PEVs with deadline $m + 1$ can be fulfilled charging. This is because we will charge the PEVs with deadline $m + 1$ first by the EDF scheduling. Thus, there exists at least a set of x_{it}'s that satisfy

$$\sum_{t=t_i^{(s)}}^{t_i^{(e)}} x_{it} = d_i, \forall i \in \{i | t_i^{(e)} = m+1\}, \tag{3.51a}$$

$$x_{i,m+1} \geq 0, \forall i \in \{i | t_i^{(e)} = m+1\}. \tag{3.51b}$$

Combining (3.45) and (3.51), we get that all the PEVs whose deadline at or before stage $m+1$ can be fulfill charging, i.e., there exist at least a set of x_{it}'s that satisfy

$$\sum_{t=t_i^{(s)}}^{t_i^{(e)}} x_{it} = d_i, \forall i \in \{i | t_i^{(e)} \leq m+1\}, \tag{3.52a}$$

$$x_{it} \geq 0, \forall t = t_i^{(s)}, \cdots, t_i^{(e)}, \forall i \in \{i | t_i^{(e)} \leq m+1\}. \tag{3.52b}$$

Therefore, we can conclude that by EDF scheduling, there always exists at least a set of x_{it}'s that is feasible to (3.1). This completes the proof. ∎

Proof of Theorem 3.2

First, we show that if there exists a PEV parking in the station at both time t_1 and t_2, i.e.,

$$t_1, t_2 \in \{t_i^{(s)}, \cdots, t_i^{(e)}\}, \tag{3.53}$$

and

$$x_{it_1}^* \geq 0, x_{it_2}^* > 0, \tag{3.54}$$

then the optimal total loads at time t_1 and t_2 must satisfy that

$$s_{t_1}^* + l_{t_1} \geq s_{t_2}^* + l_{t_2}. \tag{3.55}$$

The Karush-Kuhn-Tucker (KKT) conditions to the convex problem (3.1) are

$$f'\left(\sum_{i \in \mathscr{I}(t)} x_{it} + l_t\right) - \lambda_i - \omega_{it} = 0, i \in \mathscr{N}, t = t_i^{(s)}, \cdots, t_i^{(e)}, \tag{3.56a}$$

$$\lambda_i\left(d_i - \sum_{t=t_i^{(s)}}^{t_i^{(e)}} x_{it}\right) = 0, i \in \mathscr{N}, \tag{3.56b}$$

$$\omega_{it} x_{it} = 0, i \in \mathscr{N}, t = t_i^{(s)}, \cdots, t_i^{(e)}, \tag{3.56c}$$

where λ, ω are the non-negative optimal Lagrangian multipliers corresponding to (3.1b) and (3.1c), respectively. We separate our analysis into the following two cases:

1. If $x^*_{it_1} = 0$ for a particular PEV i at a time slot $t_1 \in \{t_i^{(s)}, \cdots, t_i^{(e)}\}$, then, by complementary slackness, we have $\omega_{it_1} > 0$. From (3.56a),

$$f'(s_{t_1} + l_{t_1}) = \lambda_i + \omega_{it_1}. \tag{3.57}$$

2. If $x^*_{it_2} > 0$ for PEV i during a time slot $t_2 \in \{t_i^{(s)}, \cdots, t_i^{(e)}\}$, we can infer from (3.56c) that $\omega_{it_2} = 0$. Then,

$$f'(s_{t_2} + l_{t_2}) = \lambda_i. \tag{3.58}$$

On the other hand, since $f(s_t + l_t)$ is a strictly convex function of $s_t + l_t$, then $f'(s_t + l_t)$ is an increasing function. From the above discussions, we get the following two conclusions:

1. If $x^*_{it_1} > 0, x^*_{it_2} > 0$, then by (3.58),

$$f'(s_{t_1} + l_{t_1}) = f'(s_{t_2} + l_{t_2}) = \lambda_i. \tag{3.59}$$

Due to the monotonicity of $f'(s_t)$, we have $s^*_{t_1} + l_{t_1} = s^*_{t_2} + l_{t_2}$.
2. If $x^*_{it_1} = 0, x^*_{it_2} > 0$, then by (3.57) and (3.58), there is

$$f'(s_{t_1} + l_{t_1}) = \lambda_i + \omega_{it_1} > f'(s_{t_2} + l_{t_2}) = \lambda_i. \tag{3.60}$$

Since $f'(s_t)$ is a increasing function, we have $s^*_{t_1} + l_{t_1} \geq s^*_{t_2} + l_{t_2}$.

Consider two function $\hat{f}(s_t + l_t)$ and $\bar{f}(s_t + l_t)$. Let $\hat{x}^*_{it}$ and $\bar{x}^*_{it}$ denote the optimal solutions to (3.1) with $f(s_t + l_t)$ replaced by $\hat{f}(s_t + l_t)$ and $\bar{f}(s_t + l_t)$, respectively. Define $\hat{s}^*_t, \bar{s}^*_t$ as

$$\hat{s}^*_t = \sum_{i \in \mathscr{I}(t)} \hat{x}^*_{it}, \bar{s}^*_t = \sum_{i \in \mathscr{I}(t)} \bar{x}^*_{it}, t = 1, \cdots, T, \tag{3.61}$$

respectively. Suppose that there exists a time slot t_1 such that

$$\hat{s}^*_{t_1} < \bar{s}^*_{t_1}. \tag{3.62}$$

Since

$$\sum_{t=1}^T \hat{s}^*_t = \sum_{t=1}^T \bar{s}^*_t = \sum_{i \in \mathscr{N}} d_i, \tag{3.63}$$

there must exist another time slot t_2 such that

$$\hat{s}^*_{t_2} > \bar{s}^*_{t_2} \tag{3.64}$$

and

$$\hat{s}_{t_1}^* + \hat{s}_{t_2}^* = \bar{s}_{t_1}^* + \bar{s}_{t_2}^* \tag{3.65}$$

Thus, we can find a PEV $i \in \mathcal{N}$ such that

$$\hat{x}_{it_1}^* < \bar{x}_{t_1}^*, \hat{x}_{it_2}^* > \bar{x}_{it_2}^*. \tag{3.66}$$

As a result,

$$\hat{x}_{it_2}^* > 0 \tag{3.67}$$

since $\bar{x}_{it_2}^* \geq 0$. Based on (3.55), there is

$$\hat{s}_{t_2}^* + l_{t_2} \leq \hat{s}_{t_1}^* + l_{t_1}. \tag{3.68}$$

Combining (3.62), (3.65), (3.68), we get

$$\bar{s}_{t_2}^* + l_{t_2} < \hat{s}_{t_2}^* + l_{t_2} \leq \hat{s}_{t_1}^* + l_{t_1} < \bar{s}_{t_1}^* + l_{t_1}. \tag{3.69}$$

Since $\bar{f}(s_t + l_t)$ is a strictly convex function of $s_t + l_t$, then, based on (3.65) and (3.69), we have

$$\bar{f}(\bar{s}_{t_1}^* + l_{t_1}) + \bar{f}(\bar{s}_{t_2}^* + l_{t_2}) > \bar{f}(\hat{s}_{t_1}^* + l_{t_1}) + \bar{f}(\hat{s}_{t_2}^* + l_{t_2}). \tag{3.70}$$

This contradicts with the fact that the $\bar{s}_t^*$ is the optimal total charging rate for objective function $\bar{f}(s_t + l_t)$. Therefore, the optimal charging solution s_t^* is the same for any strictly convex function $f(s_t + l_t)$. Next, we show that optimal solution s_t^* is a load flattening solution that minimizes $\sum_{t=1}^{T}(s_t + l_t - \frac{\sum_{t=1}^{T} s_t + l_t}{T})^2$ subjecting to (3.3b) and (3.3c). Based on the argument that s_t^* is the same for any strictly convex function $f(s_t + l_t)$, then it is equivalent to show that $\sum_{t=1}^{T}(s_t + l_t - \frac{\sum_{t=1}^{T} s_t + l_t}{T})^2$ is a strictly convex function of $s_t + l_t$. Since

$$\frac{\sum_{t=1}^{T} s_t + l_t}{T} = \frac{\sum_{i \in \mathcal{N}} d_i + \sum_{t=1}^{T} l_t}{T}, \tag{3.71}$$

which indicates that $\frac{\sum_{t=1}^{T} s_t + l_t}{T}$ is a constant. Then, we see that $\sum_{t=1}^{T}(s_t + l_t - \frac{\sum_{t=1}^{T} s_t + l_t}{T})^2$ is a strictly convex function of $s_t + l_t$. This completes the proof. ∎

Proof of Proposition 3.1

First, we show that $\Psi_1(\Xi)$ is a convex function of Ξ. For any Ξ', define $s_t^*(\Xi')$ as optimal solution that minimizes $\Psi_1(\Xi')$ subject to (3.3b)–(3.3c). Likewise, we define $s_t^*(\Xi'')$ for any Ξ''. Now let

$$\Xi''' = \lambda \Xi' + (1 - \lambda)\Xi'', \lambda \in [0, 1]. \tag{3.72}$$

Then, there must exist a feasible solution $s_t(\Xi''')$ such that

$$s_t(\Xi''') = \lambda s_t^*(\Xi') + (1 - \lambda)s_t^*(\Xi''). \tag{3.73}$$

Note that $s_t(\Xi''')$ still satisfies (3.3b)–(3.3c) due to the linearity of the constraints. Meanwhile, based on the convexity of $f(s_t + l_t)$, we have

$$\begin{aligned} \sum_{t=1}^{T} &f(s_t(\Xi''') + l_t) \\ &\leq \lambda \sum_{t=1}^{T} f(s_t^*(\Xi') + l_t) + (1 - \lambda) \sum_{t=1}^{T} f(s_t^*(\Xi'') + l_t), \end{aligned} \tag{3.74}$$

which holds for all $\lambda \in [0, 1]$. On the other hand, for $\Xi''' = \lambda \Xi' + (1 - \lambda)\Xi''$, let $s_t^*(\Xi''')$ be the optimal solution that minimizes $\sum_{t=1}^{T} f(s_t + l_t)$ under Ξ'''. Then

$$\sum_{t=1}^{T} f(s_t^*(\Xi''') + l_t) \leq \sum_{t=1}^{T} f(s_t(\Xi''') + l_t). \tag{3.75}$$

Combining (3.74) and (3.75), we have

$$\begin{aligned} \Psi_1(\Xi''') &= \sum_{t=1}^{T} f(s_t^*(\Xi''') + l_t) \\ &\leq \lambda \sum_{t=1}^{T} f(s_t^*(\Xi') + l_t) + (1 - \lambda) \sum_{t=1}^{T} f(s_t^*(\Xi'') + l_t) \\ &= \lambda \Psi_1(\Xi') + (1 - \lambda)\Psi_1(\Xi''). \end{aligned} \tag{3.76}$$

Thus, we have established the convexity of $\Psi_1(\Xi)$ over the set of Ξ. Therefore, we have

$$E[\Psi_1(\Xi)] \geq \Psi_1(E[\Xi]), \tag{3.77}$$

On the other hand, based on the definition of $s_t, \tilde{d}_t^i, \mu_t^j, i = 1, \cdots, e_1, j = t, \cdots, e_t$, we have

$$\sum_{t=1}^{T} s_t = \sum_{t=1}^{e_1} \tilde{d}_t^1 + \sum_{t=2}^{T} \sum_{j=t}^{e_t} \mu_t^j. \tag{3.78}$$

Then, by Jensen's inequality,

$$\Psi_1(E[\Xi]) = \min_{s_t} \sum_{t=1}^{T} f(s_t + v_t) \tag{3.79a}$$

$$\geq \sum_{t=1}^{T} f\left(\frac{\sum_{t=1}^{e_1} \tilde{d}_t^1 + \sum_{t=2}^{T} \sum_{j=t}^{e_t} \mu_t^j + \sum_{t=1}^{T} v_t}{T}\right) \tag{3.79b}$$

$$= Tf\left(\frac{\sum_{t=1}^{e_1} \tilde{d}_t^1 + \sum_{t=2}^{T} \sum_{j=t}^{e_t} \mu_t^j + \sum_{t=1}^{T} v_t}{T}\right). \tag{3.79c}$$

This completes the proof. ∎

Proof of Proposition 3.2

For any stage t, the following inequality holds.

$$s_t \leq \sum_{n=t}^{T} \tilde{d}_n^t. \tag{3.80}$$

Then,

$$\sum_{n=t}^{T} \tilde{d}_n^t = \sum_{n=t-1}^{T} \tilde{d}_n^{t-1} - s_{t-1} + \sum_{n=t}^{e_t} \eta_n^t \tag{3.81a}$$

$$\leq \sum_{n=t-1}^{T} \tilde{d}_n^{t-1} - \tilde{d}_{t-1}^{t-1} + \sum_{n=t}^{e_t} \eta_n^t \tag{3.81b}$$

$$= \sum_{n=t}^{T} \tilde{d}_n^{t-1} + \sum_{n=t}^{e_t} \eta_n^t \tag{3.81c}$$

$$= \sum_{m \in \{m | e_m \geq t, m=1, \cdots, t-1\}} \sum_{n=t-1}^{e_m} \eta_n^m + \sum_{n=t}^{e_t} \eta_n^t \tag{3.81d}$$

$$= \sum_{m \in \{m | e_m \geq t, m=1, \cdots, t\}} \sum_{n=t}^{e_m} \eta_n^m, \tag{3.81e}$$

where the second inequality is due to the fact that

$$s_{t-1} \geq \tilde{d}_{t-1}^{t-1}. \tag{3.82}$$

Let $\mathcal{O}(t)$ be the set that

$$\mathcal{O}(t) = \{(m,n) | e_m \geq t, m = 1, \cdots, t, n = t, \cdots, e_m\}. \tag{3.83}$$

Then we get

$$s_t \leq \sum_{(m,n) \in \mathcal{O}(t)} \eta_n^m. \tag{3.84}$$

Since $\mathcal{O}(t)$ is a bounded set for $t = 1, \cdots, T$, then

$$\mathrm{E}\left[\sum_{t=1}^{T} f\left(\sum_{(m,n) \in \mathcal{O}(t)} \eta_n^m + \iota_t\right)\right] \tag{3.85}$$

is also bounded. Thus, (3.85) is an upper bound of Φ_3. This completes the proof. ∎

Proof of Theorem 3.3

By Proposition 3.1 and Proposition 3.2, for any distribution of $\tilde{d}_n^t$ and ι_t, $t = 1, \cdots, T, n = t, \cdots, T$, we have

$$
\begin{aligned}
VSS &= \Phi_3 - \Phi_2 \\
&\leq \Phi_3 - \Phi_1 \\
&\leq \mathrm{E}\left[\sum_{t=1}^{T} f\left(\sum_{(m,n) \in \mathcal{O}(t)} \eta_n^m + \iota_t\right)\right] - Tf\left(\frac{\Gamma}{T}\right)
\end{aligned} \tag{3.86}
$$

where $\Gamma = \sum_{t=1}^{e_1} \tilde{d}_t^1 + \sum_{t=2}^{T} \sum_{j=t}^{e_t} \mu_t^j + \sum_{t=1}^{T} v_t$. This completes the proof.   ∎

Proof of Lemma 3.1

Let $\rho(i,j)$ denote the maximum density of $[i,j]$. For any $i = k, j = k + \bar{e} + 1, \cdots, T$, the density of interval $[i,j]$ is given by

$$\rho(i,j) = \frac{\sum_{k=1}^{j-k}\sum_{t=1}^{j-2k+1}\mu_t + \sum_{t=k}^{k+\bar{e}}\tilde{d}_t^k + l_k + (j-k)v}{j-k+1}$$

$$= \frac{\sum_{t=1}^{j-k}(j-k+1-t)\mu_t + \sum_{t=k}^{k+\bar{e}}\tilde{d}_t^k + l_k - v}{j-k+1} + v. \qquad (3.87)$$

To prove that the maximum density is achieved by setting $j = T$, we only need to show $\rho(i,j)$ is a non-decreasing function of j for each given i, i.e.,

$$\rho(i,j) \leq \rho(i,j+1), \forall k+\bar{e}+1 \leq j \leq T-1. \qquad (3.88)$$

Since

$$\frac{\sum_{t=1}^{j-k}(j-k+1-t)\mu_t + \sum_{t=k}^{k+\bar{e}}\tilde{d}_t^k}{j-k} \leq \sum_{t=1}^{j-k+1}\mu_t, \qquad (3.89)$$

$$k+\bar{e}+1 \leq j \leq T-1,$$

we have

$$\rho(i,j+1)$$

$$= \frac{\sum_{t=1}^{j-k}(j-k+1-t)\mu_t + \sum_{t=k}^{k+\bar{e}}\tilde{d}_t^k + \sum_{t=1}^{j-k+1}\mu_t + l_k - v}{j-k+1} + v$$

$$\geq \frac{\sum_{t=1}^{j-k}(j-k+1-t)\mu_t + \sum_{t=k}^{k+\bar{e}}\tilde{d}_t^k + l_k - v}{j-k} + v \qquad (3.90)$$

$$= \rho(i,j),$$

which implies (3.88). Hence, Y is the maximum density of $[k,j], j = k+\bar{e}+1, \cdots, T$. Next, we show that Z is the maximum density of $[k+1,T]$. For any $k+1 \leq i \leq j \leq T$, the density of interval $[i,j]$ is given by

$$\rho(i,j) = \frac{\sum_{k=1}^{j-i+1}\sum_{t=1}^{j-i+2-k}\mu_t}{j-i+1} + v$$

$$= \frac{\sum_{t=1}^{j-i+1}(j-i+2-t)\mu_t}{j-i+1} + v. \qquad (3.91)$$

To prove that the maximum density is achieved by setting $i = k+1, j = T$, we only need to show $\rho(i,j)$ is a non-decreasing function of j for each given i, i.e.,

$$\rho(i,j) \leq \rho(i,j+1), \forall k+1 \leq i \leq T-1, \qquad (3.92)$$

and a non-increasing function of i for each given j, i.e.,

$$\rho(i,j) \geq \rho(i+1,j), \forall k+1 \leq i+1 \leq j \leq T. \tag{3.93}$$

On one hand, since

$$\frac{\sum_{t=1}^{j-i+1}(j-i+2-t)\mu_t}{j-i+1} \leq \sum_{t=1}^{j-i+2}\mu_t, \forall k+1 \leq i \leq j \leq T, \tag{3.94}$$

we have

$$
\begin{aligned}
\rho(i,j+1) &= \frac{\sum_{t=1}^{j-i+1}(j-i+2-t)\mu_t + \sum_{t=1}^{j-i+2}\mu_t}{j-i+2} + v \\
&\geq \frac{\sum_{t=1}^{j-i+1}(j-i+2-t)\mu_t}{j-i+1} + v \\
&= \rho(i,j),
\end{aligned}
\tag{3.95}
$$

which implies (3.92). On the other hand, as

$$\frac{\sum_{t=1}^{j-i}(j-i+1-t)\mu_t}{j-i} \leq \sum_{t=1}^{j-i+1}\mu_t, \forall k+1 \leq i \leq j \leq T, \tag{3.96}$$

then

$$
\begin{aligned}
\rho(i+1,j) &= \frac{\sum_{t=1}^{j-i}(j-i+1-t)\mu_t}{j-i} + v \\
&\leq \frac{\sum_{t=1}^{j-i}(j-i+1-t)\mu_t + \sum_{t=1}^{j-i+1}\mu_t}{j-i+1} + v \\
&= \rho(i,j),
\end{aligned}
\tag{3.97}
$$

which implies (3.93). This completes the proof.   ∎

Proof of Lemma 3.2

First we provide the proof by discussing the following two cases:

1) If $\bar{j} \geq \bar{i}+p$, which means that $[\bar{i},\bar{j}]$ and $[\bar{i}+p,\bar{j}+p]$ overlaps with each other, then the density of period $[\bar{i},\bar{j}+p]$ is higher than that of $[\bar{i},\bar{j}]$, and the density of $[\bar{i},\bar{j}+2p]$ is higher than that of $[\bar{i},\bar{j}+p]$. So and so forth. Finally, we see that the period $[\bar{i},\bar{j}+(r-1)p]$ has the maximum density among $\{i,j|i=k+1,j=$

$i, \cdots, T$}. Thus, we have

$$\hat{i}_3 = \bar{i}, \hat{j}_3 = \bar{j} + (r-1)p, \tag{3.98}$$

and the corresponding maximum density

$$\hat{Z} = \frac{\sum_{n=\bar{i}}^{\bar{j}+(r-1)p} (\sum_{m=n}^{k+e_n} \mu_m^n + \nu_n)}{\bar{j} + (r-1)p - \bar{i} + 1}. \tag{3.99}$$

Likewise, for the region $\{i, j | i = k, j = k + \hat{e} + 1, \cdots, T\}$, we have

$$\hat{j}_2 = \bar{j} + (r-1)p, \tag{3.100}$$

and

$$\hat{Y} = \frac{\sum_{n=k}^{\bar{j}+(r-1)p} (\sum_{m=n}^{k+e_n} \mu_m^n + \nu_n)}{\bar{j} + (r-1)p - k + 1}, \tag{3.101}$$

2) If $\bar{j} < \bar{i} + p$, which means $[\bar{i}, \bar{j}], \cdots, [\bar{i} + (r-1)p, \bar{j} + (r-1)p]$ have the same maximum density among $[k+1, T]$. Then we have

$$\hat{i}_3 = \bar{i}, \hat{j}_3 = \bar{j}, \tag{3.102}$$

and the corresponding maximum density

$$\hat{Z} = \frac{\sum_{n=\bar{i}}^{\bar{j}} (\sum_{m=n}^{k+e_n} \mu_m^n + \nu_n)}{\bar{j} - \bar{i} + 1}. \tag{3.103}$$

For the region $\{i, j | i = k, j = k + \hat{e} + 1, \cdots, T\}$, if $\bar{j} \le k + \hat{e} + 1$, then $\hat{j}_2 = k + \hat{e} + 1$ since $\hat{j}_2 \ge k + \hat{e} + 1$, and

$$\hat{Y} = \frac{\sum_{n=k}^{k+\hat{e}+1} (\sum_{m=n}^{k+e_n} \mu_m^n + \nu_n)}{\hat{e} + 2}. \tag{3.104}$$

Otherwise, $\hat{j}_2 = \bar{j}$, and

$$\hat{Y} = \frac{\sum_{n=k}^{\bar{j}} (\sum_{m=n}^{k+e_n} \mu_m^n + \nu_n)}{\bar{j} - k + 1}. \tag{3.105}$$

This completes the proof. ∎

References

1. W. Tang and Y. J. Zhang, "A model predictive control approach for low-complexity electric vehicle charging scheduling: optimality and scalability," accepted by *IEEE Trans. Power Syst.*, Jun. 2016.
2. Z. Ma, D. Callaway, and I. Hiskens, "Decentralized Charging Control of Large Populations of Plug-in Electric Vehicles", *IEEE Trans. on Control Systems Technology*, vol.21, no.1, pp. 67–78, 2013.
3. Y. He, B. Venkatesh, and L. Guan, "Optimal scheduling for charging and discharging of electric vehicles," *IEEE Trans. on Smart Grid*, vol.3, no.3, pp. 1095–1105, 2012.
4. N. Chen, L. Gan, S. H. Low, and A. Wierman, "Distributional analysis for model predictive deferrable load control," *53rd IEEE Conference on Decision and Control (CDC)*, pp. 6433–6438, Dec. 2014.
5. L. Gan, A. Wierman, U. Topcu, N. Chen, and S. H. Low, "Realtime deferrable load control: handling the uncertainties of renewable generation," *in Proceedings of the fourth international conference on Future energy systems (ACM e-Energy)* , pp. 113–124, May 2013.
6. A. Shapiro, D. Dentcheva, and A. Ruszczynski, *Lectures on Stochastic Programming: Modeling and Theory*, MPS-SIAM, Philadelphia, 2009.
7. J. R. Birge and F. Louveaux, *Introduction to Stochastic Programming*, New York: Springer, 1997.
8. B. Defourny, D. Ernst, L. Wehenkel, L. E. Sucar, E. F. Morales, and J. Hoey, "Multistage stochastic programming: A scenariotree based approach to planning under uncertainty," *in DecisionTheory Models for Applications in Artificial Intelligence: Concepts and Solutions*, pp. 51, 2011.
9. F. Maggioni and S. Wallace, "Analyzing the quality of the expected value solution in stochastic programming," *Annals of Operations Research*, pp. 37–54, 2012.
10. R. Leou, C. Su, and C. Lu, "Stochastic analyses of electric vehicle charging impacts on distribution network," *IEEE Trans. on Power System*, vol. 29, no. 3, pp. 1055–1063, 2014.
11. L. Rao and J. Yao, "SmartCar: smart charging and driving control for electric vehicles in the smart grid," *IEEE Global Communications Conference (GLOBECOM)*, pp. 2709–2714, Dec. 2014.
12. S. Bansal, M. N. Zeilinger, and C. J. Tomlin, "Plug-and-play model predictive control for electric vehicle charging and voltage control in smart grids," *53rd IEEE Conference on Decision and Control (CDC)*, pp. 5894–5900, 2014.
13. Y. Ye, *Interior Point Algorithms: Theory and Analysis*, Wiley-Interscience Press, 1997.
14. F. Yao, A. Demers, and S. Shenker, "A scheduling model for reduced cpu energy," *in Proc. IEEE Symp. Foundations of Computer Science*, pp. 374–382, 1995.
15. N. Bansal, T. Kimbrel, and K. Pruhs, "Speed scaling to manage energy and temperature," *Journal of the ACM (JACM)*, vol. 54, no. 1, pp. 1–39, 2007.
16. W. Tang, S. Bi, and Y. J. Zhang, "Online coordinated charging decision algorithm for electric vehicles without future information," *IEEE Trans. on Smart Grid*, vol. 5, no. 6, pp. 2810–2824, 2014.
17. A. Santos, A. N. McGuckin, H. Y. Nakamoto, D. Gray, and S. Lis, *Summary of travel trends: 2009 national household travel survey*, Federal Highway Administration, Washington, DC, 2011.
18. Southern california edison dynamic load profiles, https://www.sce.com/wps/portal/home/regulatory/load-profiles, 2012.

Chapter 4
Optimal BESS Control in Microgrids

With the wide penetration of the renewable energy generators in power grids, the high variability of renewable source output poses significant challenges to the power grid, including voltage instability and power generation cost. Utilising BESS in microgrids is considered an effective mechanism for absorbing the fluctuation of local energy generation and consumption, and thereby mitigating the detrimental impact of renewable energy sources on the main grid. In this chapter, we consider the optimal control of a microgrid, e.g., a factory or a commercial building, which is equipped with a finite-capacity BESS and renewable energy generators. The microgrid tries to meet the demand using the power drawn from the renewable generator, BESS, as well as the main grid. In particular, the optimal control decisions are made without the non-causal knowledge of future load demand and renewable source output. We aim to minimize a cost function, which is a general convex increasing function of instantaneous power flow from the grid. The increasing convexity of the cost function ensures that the power flow exchanged with the main grid remains as flat over time as possible, and thus minimizes the negative impact of renewable energy integration.

4.1 Online Optimal Battery Charging Problem

We consider a microgrid, e.g., a factory or a commercial building, which is equipped with a finite-capacity BESS and renewable energy generators. The microgrid tries to meet the demand using the power drawn from the renewable generator, BESS, as well as the main grid.

Suppose that the entire system time is divided into T equal-length time slots. Suppose that the length of the time slots is normalized to 1. We denote by $E(t)$ the amount of energy stored in the battery at the beginning of time slot t, and by ζ the

© The Author(s) 2017
W. Tang, Y.J.A. Zhang, *Optimal Charging Control of Electric Vehicles in Smart Grids*, SpringerBriefs in Electrical and Computer Engineering,
DOI 10.1007/978-3-319-45862-5_4

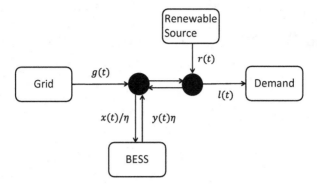

Fig. 4.1 The illustration of the system model

energy capacity of the battery. Then, the energy stored in the battery should satisfy

$$0 \le E(t) \le \zeta \tag{4.1}$$

for all t. Let $x(t)$ and $y(t)$ be the charging and discharging power during time slot t, respectively. Notice that since the length of a slot is normalized to 1, $x(t)$ and $y(t)$ are also the energy charged or discharged during the time slot. The power that can be charged or discharged from the battery is always bounded. Let us denote by u_x the maximum charging power and u_y the maximum discharging power. Then, $x(t), y(t)$ should satisfy

$$0 \le x(t) \le u_x, 0 \le y(t) \le u_y, \tag{4.2}$$

respectively. Note that the battery cannot be charged and discharged simultaneously. Therefore, $x(t)$ and $y(t)$ cannot be non-zero at the same time. The battery has the following dynamics:

$$E(t+1) = E(t) + x(t) - y(t). \tag{4.3}$$

Since the battery is operated on dc, an ac-to-dc converter or a dc-to-ac converter is necessary when the battery is being charged or discharged. For simplicity, suppose that both converters have the same constant conversion efficiency[1] η, where

$$0 < \eta \le 1. \tag{4.4}$$

As a result, the ac power drawn from (or injected into) the microgrid is $x(t)/\eta$ (or $\eta y(t)$) when BESS is charged (or discharged). Likewise, let $l(t)$ be the load demand

---

[1] In general, the charging and discharging efficiency may not be the same. The result in this chapter can be easily extended to the case when the charging and discharging efficiency are different.

of the microgrid at time t. Suppose that there exists a renewable energy source, e.g., solar energy, wind power or some biomass materials, which is connected to the ac bus. Let $r(t)$ be the power generated by the renewable sources at time t. Now we define an auxiliary variable $p(t)$,

$$p(t) = l(t) - r(t), t = 1, \cdots, T, \tag{4.5}$$

which denotes the net difference between the load demand and renewable power generated at time t. The space of $p(t)$ is denoted by $\mathscr{P}$. We denote by $g(t)$ the power purchased from (i.e., $g(t) > 0$) or sold to (i.e., $g(t) < 0$) the main grid at time t, as shown in Fig. 4.1. To balance the power supply and demand in the microgrid, $g(t)$ is calculated as

$$g(t) = \frac{x(t)}{\eta} - \eta y(t) + p(t). \tag{4.6}$$

We assume that the instantaneous cost associated with the power flow from the grid $g(t)$ is an convex increasing function, denoted as $f(g(t))$. On one hand, the convexity of $f(.)$ reflects the fact that each unit of additional power demand becomes more expensive to obtain and make available to the consumer. For example, in the wholesale market, the instantaneous cost can be modeled as a quadratic function of the instant load [1–3]. On the other hand, it also captures the intent of flattening $g(t)$ over time [4], so that the detrimental effect on the main grid is reduced.

The *system state* is defined as $(E(t), p(t))$ at each time t. Specifically, $E(t)$ is defined as the endogenous state variable that depends on the decision variable $x(t), y(t)$ through the transition rules (4.3), and $p(t)$ is defined as the exogenous state variable that does not depend on the previous decisions [5]. Let $\mathscr{E}$ be the variation space of the $E(t)$, i.e.,

$$\mathscr{E} = \{E(t)|0 \le E(t) \le \zeta\}. \tag{4.7}$$

Then, the state space is given by $\mathscr{E} \times \mathscr{P}$. Given the system state $(E(t), p(t))$, the corresponding action space, denoted by $\mathscr{S}(E(t))$, is given by

$$\begin{aligned} \mathscr{S}(E(t)) = \{x(t), y(t)|0 \le x(t) \le \min\{\zeta - E(t), u_x\}, \\ 0 \le y(t) \le \min\{E(t), u_y\}\}. \end{aligned} \tag{4.8}$$

Our objective is to calculate, at each time t, the optimal battery charging policy without knowing the realization of the future load $l(t)$ and the renewable source output $r(t)$. In other words, at each time t, we solve the following problem to find the optimal policy $x^*(t)$ and $y^*(t)$ that achieves the optimal cost-to-go $G_t^*(E(t), p(t))$ at current state $(E(t), p(t))$.

$$G_t^*(E(t), p(t)) \tag{4.9a}$$

$$= \min_{x(t),y(t)} f(g(t)) + \lambda \mathbf{E}_{p(t+1) \in \mathscr{P}}[G_{t+1}^*(E(t+1),p(t+1))] \tag{4.9b}$$

$$\text{s. t. } g(t) = \frac{x(t)}{\eta} - \eta y(t) + p(t), \tag{4.9c}$$

$$E(t+1) = E(t) + x(t) - y(t), \tag{4.9d}$$

$$x(t), y(t) \in \mathscr{S}(E(t)), t = 1, \cdots, T, \tag{4.9e}$$

where λ is a discounted factor, which represents the value reduction over time. The following proposition prove that there always exists an optimal solution where $x^*(t)$ and $y^*(t)$ are not nonzero at the same time. That is, the battery is not charged and discharged at the same time under the optimal solution, even though such a constraint is not included in (4.9) explicitly.

4.2   Related Work

In this chapter, we rigorously prove that the optimal BESS operation policy exhibits a threshold structure, which can potentially lead to a simplified control policy. Specifically, we show that the optimal policy degenerates to one that displays *short-sighted* behaviour when the discount factor satisfies certain conditions. Moreover, we show that the optimal cost is a decreasing convex function of the battery capacity, implying that there exists an optimal battery sizing that strikes a balance between the operation cost and the capital investment. Our analyses are validated through extensive simulations. Our results demonstrate that BESS can significantly reduce the fluctuation of power flow exchanged with the main grid.

Optimal BESS control mechanisms have been recently studied for systems with renewable energy sources [6–12]. For simplicity, most of the work assumed that the cost function is linear with the instantaneous power drawn from the main grid [6–9]. Therein, BESS was mainly used for energy arbitrage by exploiting the electricity price variation in the main grid. On the contrary, our chapter aims to reduce the fluctuation in the power flow exchanged with the main grid. The flattened power flow greatly reduces the vulnerability of the power system, thus reducing the negative impact of renewable energy source integration. To this end, we adopt an increasing convex cost function instead of a linear one. Similar convex cost functions were also considered in [10, 11]. Reference [10] devises an online dynamic control policy to minimize long-term average grid operational cost. Meanwhile, Lyapunov optimization technique was used in [11, 12] to minimize the long term average cost of a power consuming entity. Note that the control polices provided in [10–12] are suboptimal. For instance, the control polices provided in [10, 11] are asymptotically optimal as the storage capacity becomes large. That is, the gap from optimality increases as storage capacity decreases. Whereas in this chapter, the control policy is strictly optimal for any battery capacity, and the optimality is not affected by the battery capacity. Besides, [10] does not take into account the charging efficiency of the battery converter and the upper bounds on the charging and discharging rates. In contrast, both the battery charging efficiency and the maximum charging/discharging rate have been considered in this chapter.

4.3 Optimal Charging Policy

In this section, we derive the optimal charging policy at each time slot t depending on BESS energy level $E(t)$ as well as $l(t)$ and $r(t)$ in that time slot. First, we introduce the following proposition which shows one feature of the optimal charging solution.

Proposition 4.1. *There always exists an optimal solution $x^*(t), y^*(t)$ to (4.9) such that $x^*(t)$ and $y^*(t)$ are not nonzero at the same time.*

Proof. Please see the detailed proof in Appendix "Proof of Proposition 4.1".

Proposition 4.1 provides a class of optimal policy which is relatively easy to calculated. In the next sections, we always calculate the optimal policy $x^*(t), y^*(t)$ where $x^*(t)$ and $y^*(t)$ cannot be nonzero at the same time. Due to Proposition 4.1, it is safe to ignore the constraint that BESS cannot be charged and discharged at the same time in (4.9), knowing that such a constraint is automatically satisfied in the remaining of this chapter, we thus focus on the optimal policy $x^*(t)$ and $y^*(t)$ where they are not positive at the same time.

4.3.1 Optimal Battery Charging Policy by Value Iteration

In practice, the length of each time slot is sufficiently small compared with the entire system time T, so that T can be viewed as infinite compared with t. In this case, Problem (4.9) can be treated as an infinite-time horizon dynamic programming and the time index in (4.9) can be removed. We use E, p, x^*, y^* and $G^*(E,p)$ to replace $E(t), p(t), x^*(t), y^*(t), G^*(E(t), p(t))$ at time t, respectively. We define an auxiliary function $\hat{G}(E,p) \forall x, y \in \mathscr{S}(E)$,

$$\tilde{G}(E,p,x,y) = f(\frac{x}{\eta} - \eta y + p) + \lambda \mathbf{E}_{q \in \mathscr{P}}[G^*(E+x-y,q)]. \tag{4.10}$$

Then, $G^*(E,p)$ satisfies the Bellman's equation

$$G^*(E,p) = \min_{x,y \in \mathscr{S}(E)} \tilde{G}(E,p,x,y). \tag{4.11}$$

where x^* and y^* are the optimal solutions to Problem (4.11). x^* and y^* can be derived through the standard value iteration algorithm [13], as shown in Algorithm 3, where ε is defined as the optimality gap between the cost-to-go value output by Algorithm 3 and the optimal cost-to-go value.

Algorithm 3: Optimal battery charging algorithm by value iteration

---

    input : $\mathscr{E}, \mathscr{P}, \varepsilon, \lambda$
    output: $G^{n+1}(E,p), x^*, y^*$
1 initialization $n = 0, G^0(E,p) = 0 \ \forall E \in \mathscr{E}, p \in \mathscr{P}$;
2 **repeat**
3   For $E \in \mathscr{E}, p \in \mathscr{P}$, compute $G^{n+1}(E,p)$ by

$$G^{n+1}(E,p)$$
$$= \min_{x,y \in \mathscr{S}(E)} f(\frac{x}{\eta} - \eta y + p) + \lambda \mathbf{E}_{q \in \mathscr{P}}[G^n(E + x - y, q)]. \tag{4.12}$$

4 **until** $||G^{n+1} - G^n|| \leq \varepsilon(1 - \lambda)/(2\lambda)$;
5 For $E \in \mathscr{E}, p \in \mathscr{P}$, compute x^*, y^* by

$$x^*, y^* \in \arg \min_{x,y \in \mathscr{S}(E)} f\left(\frac{x}{\eta} - \eta y + p\right) + \lambda \mathbf{E}_{q \in \mathscr{P}}[G^{n+1}(E + x - y, q)]. \tag{4.13}$$

4.3.2 Threshold Structure of the Optimal Policy

In this subsection, we characterize a two-threshold structure property of the optimal policy. First, we introduce the following Lemma that helps prove the threshold structure.

Lemma 4.1. *The cost function $G^*(E,p)$ and $\mathbf{E}_{p \in \mathscr{P}}[G^*(E,p)]$ are convex functions of in $E \in \mathscr{E}$ for each given $p \in \mathscr{P}$.*

Proof. Please see the detailed proof in Appendix "Proof of Lemma 4.1".

Based on Lemma 4.1, we establish the threshold-based structure of the optimal policy for Problem (4.11) in Theorem 4.1.

Theorem 4.1. *For each state (E,p), there exists two threshold boundaries, denoted by $\xi_1(E,p)$ and $\xi_2(E,p)$, where $\xi_1(E,p), \xi_2(E,p) \in \mathscr{E}$ and $\xi_1(E,p) \leq \xi_2(E,p)$. The optimal battery charging policy for the state (E,p) is given by*

$$(x^*, y^*) =$$

$$\begin{cases} (\min\{\min\{\zeta, \xi_1(E,p)\} - E, u_x\}, 0) & \text{if } E \leq \xi_1(E,p), \\ (0, \min\{E - \max\{0, \xi_2(E,p)\}, u_y\}) & \text{if } E \geq \xi_2(E,p), \\ (0,0) & \text{otherwise.} \end{cases} \tag{4.14}$$

Proof. Please see the detailed proof in Appendix "Proof of Theorem 4.1".

Theorem 4.1 states the fact that if the current energy level E is smaller than $\xi_1(E,p)$, then optimal policy is to increase the energy level to as close to $\xi_1(E,p)$ as possible. If E is larger than $\xi_2(E,p)$, then optimal policy is to decrease the energy level to as close to $\xi_2(E,p)$ as possible. Otherwise, the optimal policy is to neither charge or discharge the battery.

4.4 Degeneration to a Short-Sighted Policy

If the future is not considered at each time slot, i.e., λ in (4.9) is set to be 0, then the optimal policy takes a *short-sighted* behavior that discharges the battery as fast as possible for any state. This is because $G^*(E,p)$ only minimizes the function $f(x/\eta - \eta y + p)$ when $\lambda = 0$. This shortsighted policy drains the battery as fast as possible without recharging it, making BESS almost useless in the system operation. When $0 < \lambda < 1$, it is still possible that the optimal policy degenerates to the short-sighted policy. In this section, we show that if λ is no larger than the *valley-peak ratio*, defined as the ratio of the lowest possible total load to the highest possible total load, then the optimal policy is to discharge the battery as fast as possible for all states. Note that the valley-peak ratio may not be small in practice. For example, the valley-peak ratio is calculated to be 0.3408 by adopting the real data of the load and the system parameters in Sect. 4.6.3. In this case, to avoid the short-sighted policy, λ should be set to be more larger than 0.3408.

For all $p \in \mathscr{P}$, let p_1, p_2 be the lower bound and upper bound of p, respectively. Now we are ready to present Proposition 4.2.

Proposition 4.2. *For all $\eta \in (0,1]$, if $\lambda, p_1, p_2, u_x, u_y$ satisfy the follow inequality,*

$$p_1 \geq u_y, \lambda \leq \frac{p_1 - \eta u_y}{p_2 + \frac{u_x}{\eta}}, \tag{4.15}$$

then, the thresholds satisfy

$$\xi_1(E,p) \leq \xi_2(E,p) \leq \max\{0, E - u_y\}, \tag{4.16}$$

and the charging policy degenerates to

$$x^* = 0, y^* = \min\{E, u_y\}. \tag{4.17}$$

Proof. Please see the detailed proof in Appendix "Proof of Proposition 4.2".

4.5 Optimal Battery Sizing

Intuitively, a battery with larger capacity can absorb higher variability and achieve a lower total cost. In this section, we are interested to find the optimal investment in battery capacity without non-causally knowing the future load and renewable sources. Specifically, for each given state (E,p), we treat the optimal cost-to-go $G^*(E,p)$ as a function of ζ and analyze how $G^*(E,p)$ changes as ζ changes. That is, for each given $\zeta, \zeta \geq 0$, First, by Theorem 4.2, we show that $G^*(E,p)$ is a decreasing convex function of ζ.

Theorem 4.2. *The optimal value $G^*(E,p)$ is a decreasing convex function of ζ.*

Proof. Please see the detailed proof in Appendix "Proof of Proposition 4.2".

Theorem 4.2 leads to a fact that the optimal cost-to-go decreases quickly as the battery capacity increases at first and then decreases more and more slowly as battery capacity increases. There exists an optimal battery sizing that strikes a balance between the operation cost and the capital investment. The numerical results in Sect. 4.6.2 also verifies this conclusion.

4.6 Simulations

In this section, we will illustrate our theoretical results through extensive numerical simulations. Similar to [1, 3], we adopt a quadratic cost function in the simulations, i.e.,

$$f(g(t)) = g^2(t). \tag{4.18}$$

4.6.1 The Optimal Policy in General Case

We use Algorithm 3 provided in Sect. 4.3.1 to compute the optimal charging policy and validate the threshold structure. The length of each time slot is set to be one hour. The battery in our simulation is chosen from [14] with maximum charging/discharging rate 3 kW, battery capacity 35 kWh. Based on the typical

Fig. 4.2 Optimal policy over
system state for the battery
with $\zeta = 35$ kWh

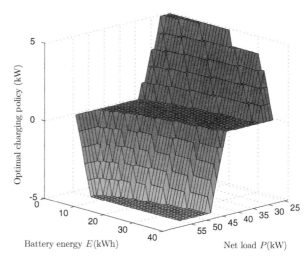

parameter setting of the sodium sulfur (NaS) battery [15], we set the conversion
efficiency $\eta = 0.85$. The discount factor is set to be $\lambda = 0.9$ and the optimal gap
ε is set to be $\varepsilon = 0.001$. According to the load profile of the service area in South
California Edison during Feb. 2011 [1], the minimum and maximum of base load
are given by

$$p_1 = 18\text{kW}, p_2 = 46\text{kW}, \tag{4.19}$$

respectively. Then, the base load (kW) is assumed to be uniformly distributed in
$[18, 46]$. We adopt the wind energy source as the renewable source and assume that
the wind power (kW) is set to be uniformly distributed in $[7, 9]$. Similar settings
of wind power have been adopted in [11]. We consider two types of batteries
[14]: (1) maximum charging/discharging rate $u_x = u_y = 5$ kW, battery capacity
$\zeta = 35$ kWh; (2) maximum charging/discharging rate $u_x = u_y = 5$ kW, battery
capacity $\zeta = 10$ kWh. For all batteries, we discretize the continuous spaces of the
battery energy level and charging/discharging rate, where the step sizes are set to be
1 kWh and 1kW, respectively.

We plot the optimal policy versus system state for battery with capacity 35 kWh
in Fig. 4.2. An observation is that when both the net load p and the battery energy
level E are very high, the optimal policy tends to discharge the battery as fast as
possible. When both the net load p and the battery energy level E are very low, the
optimal policy tends to charge the battery as fast as possible. When the net load p
and the battery energy level E are neither too high nor too low, it is better to not
charge or discharge the battery. As such, the optimal policy efficiently avoids the
extremely high or low level of total load g.

To further verify that the battery can efficiently reduce the fluctuation caused
from the renewable energy, we randomly generate the sample of base load and wind
power within 50 time slots, shown as the black line with square in Fig. 4.3. For both

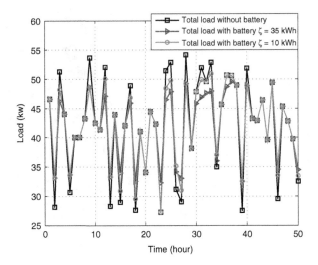

Fig. 4.3 Total load comparison among different type of batteries

batteries, we set the initial battery energy level to be half of the battery capacity. Then, from time slot 1 to 50, we calculate the total loads of both batteries at each time slot based on their optimal charging policies, respectively. We draw the curve of the total load of two batteries under the optimal charging policy, respectively, as shown in Fig. 4.3. Figure 4.3 shows that both two type of batteries can efficiently reduce the total load at the peak time, e.g., 24, 25, 30–33 and raise the total load at the valley time, e.g., 18, 23, 26, 27. Besides, we see that the battery with $\zeta = 35$ kWh has more ability to absorb the fluctuations than the battery with $\zeta = 10$ kWh. Therefore, we see that under the optimal charging policy, BESS flattens the power flow from the main grid as much as possible.

4.6.2 The Optimal Battery Capacity

In this subsection, we investigate how the battery capacity influences the total cost under the optimal policy in practical scenario. We let the capacity of the battery ζ increase from 0 to 40 kWh. Other parameter settings are the same as Sect. 4.6.1. For each battery capacity ζ, we calculate the average cost of the optimal cost-to-go over all system states under the optimal policy. In order to facilitate comparison, we normalize all the average costs over the average cost without battery, i.e., $\zeta = 0$ kWh. We plot the normalized cost over the battery capacity in Fig. 4.4. Figure 4.4 shows that the normalized cost under the optimal policy is a decreasing convex function of the battery capacity ζ, which is consistent with Theorem 4.2 in Sect. 4.5. We see that when the battery capacity increases from 0 to 40 kWh, the normalized cost decreases from 1 to 0.7848. That is, the microgrid with a battery

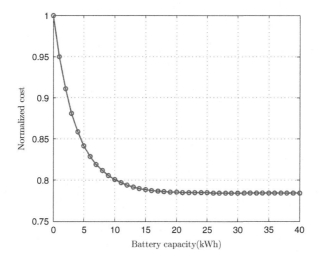

Fig. 4.4 The normalized cost via the battery capacity

capacity 24 kWh saves 21.52 % cost compared with that without battery. When the battery capacity exceeds 24 kWh, the normalized cost decreases very slightly as the battery capacity increases. Thus, to achieve the most economic benefit, 24 kWh is suggested to be the best choice of the battery capacity.

4.6.3 The Short-Sighted Policy Under Special λ

In this subsection, we simulate the short-sighted policy by setting λ to be no larger than the valley-peak ratio. The empirical distribution of net load and the settings of the batteries are the same as Sect. 4.6.1. Then, p_1, u_y satisfies

$$p_1 \geq u_y. \tag{4.20}$$

The valley-peak ratio is calculated as

$$\frac{p_1 - \eta u_y}{p_2 + u_x/\eta} = 0.3408. \tag{4.21}$$

To ensure that the condition of Proposition 4.2 is satisfied, we set the discount factor λ to be 0.3408. Besides, we set the optimal gap $\varepsilon = 0.001$. We still use Algorithm 3 provided in Sect. 4.3.1 to compute the optimal charging policy with $\lambda = 0.3408$. Figure 4.5 shows the optimal charging policy of the battery with $\zeta = 35$ kWh. We see that the optimal charging policy always discharges the battery as fast as possible for all states. Similar to Sect. 4.6.1, we simulate the performance of fluctuation

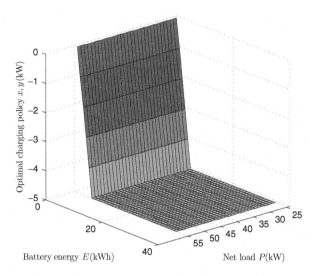

Fig. 4.5 The short-sighted policy over system state for the battery with $\zeta = 35$ kWh

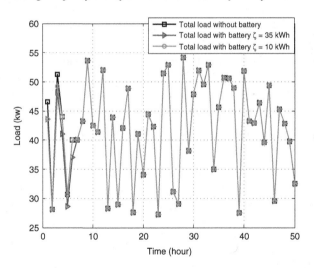

Fig. 4.6 Total load comparison among different type of batteries under the short-sighted policy with $\lambda = 0.3408$

reduction under current optimal charging policy. The samples of base load and net load are adopted the same as Sect. 4.6.1. We draw the curve of the total loads with two batteries under current optimal policy as well as the total load without battery, as shown in Fig. 4.6. Figure 4.6 indicates that the charging policy takes a short-sighted behaviour that can hardly reduce the load fluctuation since it discharges two batteries as soon as possible at the first 7 hours and then never charge or discharge again. Compared with Fig. 4.3, Fig. 4.6 shows that for both two types of batteries, the short-sighted optimal policy with $\lambda = 0.3408$ performs much worse than the optimal policy with $\lambda = 0.9$ in the matter of flatten the total load exchanged with the main grid. Therefore, to avoid the short-sighted policy, λ is suggested to be much more larger than 0.3408.

4.7 Conclusions

This chapter investigates the problem of optimal BESS control in a microgrid with renewable energy sources. In particular, we derived an optimal BESS operation policy that minimizes a cost function, which is a general convex increasing function of instantaneous power flow from the grid. The increasing convexity of the cost function makes sure that the power flow from the main grid is as flat over time as possible. Through rigorous analysis, we show that the optimal policy exhibits a threshold structure. Specially, we prove that the optimal policy take a *short-sighted* behavior when the discount factor satisfies certain conditions. Additionally, we show that the optimal cost is a decreasing convex function of the battery capacity, which implies that there exists an optimal battery sizing that strikes a balance between the total cost and the capital investment. The simulations validate the analyses and show that BESS can significantly reduce the fluctuation of power flow exchanged with the main grid.

Appendix

Proof of Proposition 4.1

Note that $0 < \eta \leq 1$ by definition. We first prove the proposition for the case when $0 < \eta < 1$ by contradiction. Let $x'(t) > 0, y'(t) > 0$ be an optimal solution to (4.9). If $x'(t) \geq y'(t)$, then, there exists another solution $x''(t), y''(t)$ such that $x''(t) = x'(t) - y'(t) \geq 0, y''(t) = 0$, which satisfies the constraint (4.9d)–(4.9e). Since $0 < \eta < 1$, the following inequality always holds.

$$x'(t)/\eta - \eta y'(t) + l(t) - r(t) > x''(t)/\eta - \eta y''(t) + l(t) - r(t). \qquad (4.22)$$

Note that $E(t) + x'(t) - y'(t) = E(t) + x''(t) - y''(t)$. Thus,

$$\mathbf{E}\left[\lim_{T\to\infty}\sum_{t=1}^{T}\lambda^t f\left(\frac{x'(t)}{\eta}-\eta y'(t)+l(t)-r(t)\right)\right]$$

$$>\mathbf{E}\left[\lim_{T\to\infty}\sum_{t=1}^{T}\lambda^t f\left(\frac{x''(t)}{\eta}-\eta y''(t)+l(t)-r(t)\right)\right],$$

(4.23)

which contradicts the assumption that $x'(t) > 0, y'(t) > 0$ is the optimal solution. Similarly, we can show that $x'(t) > 0, y'(t) > 0$ cannot be the optimal solution when $x'(t) < y'(t)$. Next, we prove the proposition when $\eta = 1$. In this case, the variables $x(t), y(t)$ in (4.9) can be replaced by a single variable $h(t)$, where

$$h(t) = x(t) - y(t), -u_y \le h(t) \le u_x.$$

(4.24)

Let us denote by $h^*(t)$ the optimal solution to (4.9). If $h^*(t) \le 0$, the $x^* = 0, y^*(t) = -h^*(t)$ is one of the optimal solutions to (4.9). If $h^*(t) > 0$, $x^* = h^*(t), y^*(t) = 0$ is one of the optimal solutions to (4.9). This completes the proof. ∎

Proof of Lemma 4.1

We show the lemma by induction. Define $G^k(E,p)$ as the cost-to-go value at kth value iteration in Algorithm 3. First, when $k = 0$,

$$G^0(E,p) = 0 \; \forall E \in \mathscr{E}, p \in \mathscr{P}$$

(4.25)

and

$$\mathbf{E}_{p\in\mathscr{P}}[G^0(E,p)] = 0 \; \forall E \in \mathscr{E}.$$

(4.26)

Thus, it is trivial to see that $G^0(E,p)$ is convex in $E \in \mathscr{E} \; \forall p \in \mathscr{P}$. Now we show that if $G^k(E,p)$ is convex in $E \in \mathscr{E} \; \forall p \in \mathscr{P}$, then $G^{k+1}(E,p)$ is also convex in $E \in \mathscr{E}$ $\forall p \in \mathscr{P}$. Suppose that $G^k(E,p)$ is convex in $E \in \mathscr{E} \; \forall p \in \mathscr{P}$. Then, $\mathbf{E}_{p\in\mathscr{P}}[G^k(E,p)]$ is convex in $E \in \mathscr{E}$. At $(k+1)$th value iteration, for any $E \in \mathscr{E}$,

$$G^{k+1}(E,p)$$
$$= \min_{x,y\in\mathscr{S}(E)} f\left(\frac{x}{\eta}-\eta y+p\right) + \lambda\mathbf{E}_{q\in\mathscr{P}}[G^k(E+x-y,q)].$$

(4.27)

For any $E_1 \in \mathscr{E}$, define $x^*(E_1), y^*(E_1)$ as optimal solution to (4.27) with $E = E_1$. That is,

$$G^{k+1}(E_1,p) = f\left(\frac{x^*(E_1)}{\eta} - \eta y^*(E_1) + p\right) \tag{4.28}$$
$$+ \lambda \mathbf{E}_{q \in \mathscr{P}}[G^k(E_1 + x^*(E_1) - y^*(E_1),q)].$$

Likewise, we define $x^*(E_2), y^*(E_2)$ for any $E_2 \in \mathscr{E}$. Now let

$$E_3 = \alpha E_1 + (1-\alpha)E_2. \tag{4.29}$$

Note that E_3 satisfies that $E_3 \in \mathscr{E}$. Then, there must exist $x_t(E_3), y_t(E_3)$ such that

$$x(E_3) = \alpha x^*(E_1) + (1-\alpha)x^*(E_2),$$
$$y(E_3) = \alpha y^*(E_1) + (1-\alpha)y^*(E_2). \tag{4.30}$$

Because $x^*(E_1), y^*(E_1) \in \mathscr{S}(E_1)$ and $x^*(E_2), y^*(E_2) \in \mathscr{S}(E_2)$, we have $x(E_3), y(E_3) \in \mathscr{S}(E_3)$ due to the linearity of (4.8). Since

$$E_3 + x(E_3) - y(E_3)$$
$$= \alpha(E_1 + x^*(E_1) - y^*(E_1)) + (1-\alpha)(E_2 + x^*(E_2) - y^*(E_2)), \tag{4.31}$$

where $E_1 + x^*(E_1) - y^*(E_1), E_2 + x^*(E_2) - y^*(E_2) \in \mathscr{E}$. Then, we have $E_3 + x(E_3) - y(E_3) \in \mathscr{E}$. Let $x^*(E_3), y^*(E_3)$ be the optimal solution to (4.27) with $E = E_3$. Then, for all $p \in \mathscr{P}$,

$$G^{k+1}(E_3,p)$$
$$= f\left(\frac{x^*(E_3)}{\eta} - \eta y^*(E_3) + p\right)$$
$$+ \lambda \mathbf{E}_{q \in \mathscr{P}}[G^k(E_3 + x^*(E_3) - y^*(E_3),q)] \tag{4.32a}$$
$$\leq f(x(E_3)/\eta - \eta y(E_3) + p) + \lambda \mathbf{E}_{q \in \mathscr{P}}[G^k(E_3 + x(E_3) - y(E_3),q)] \tag{4.32b}$$
$$\leq \alpha f(x^*(E_1)/\eta - \eta y^*(E_1) + p) + (1-\alpha)f(x^*(E_2)/\eta - \eta y^*(E_2)$$
$$+ p) + \alpha \lambda \mathbf{E}_{q \in \mathscr{P}}[G^k(E_1 + x^*(E_1) - y^*(E_1),q)]$$
$$+ (1-\alpha)\lambda \mathbf{E}_{q \in \mathscr{P}}[G^k(E_2 + x^*(E_2) - y^*(E_2),q)] \tag{4.32c}$$
$$= \alpha G^{k+1}(E_1,p) + (1-\alpha)G^{k+1}(E_2,p), \tag{4.32d}$$

where (4.32a) and (4.32b) hold based on the definition of $x^*(E_3), y^*(E_3)$ and $x(E_3), y(E_3)$, (4.32b) holds based on the definition of $x^*(E_3), y^*(E_3)$ and $x(E_3), y(E_3)$, (4.32c) holds due to the convexity of $f(\frac{x(E_3)}{\eta} - \eta y(E_3) + p)$ and $\lambda \mathbf{E}_{q \in \mathscr{P}}[G^k(E_3 + x(E_3) - y(E_3),q)]$, and (4.32d) holds based on the definition of $x^*(E_1), y^*(E_1)$ and $x^*(E_2), y^*(E_2)$. Thus, $G^{k+1}(E,p)$ is convex in $E \in \mathscr{E}$ for all $p \in \mathscr{P}$. Therefore, we have shown that $G^k(E,p)$ is convex in $E \in \mathscr{E}$ for all $p \in \mathscr{P}, k = 0, 1, 2, \cdots$. Similarly, we can show that $\mathbf{E}_{p \in \mathscr{P}}[G^k(E,p)]$ is convex in $E \in \mathscr{E}$ for each given $p \in \mathscr{P}$. Since

$$\lim_{k \to \infty} G^k(E,p) = G^*(E,p) \; \forall E \in \mathscr{E}, p \in \mathscr{P}, \tag{4.33}$$

it is suffices to show that $G^*(E,p)$ is convex in $E \in \mathscr{E}$ for all $p \in \mathscr{P}$. Due to the linearity of the expectation operator, we see that $\mathbf{E}_{p \in \mathscr{P}}[G^*(E,p)]$ is convex in $E \in \mathscr{E}$ for all $p \in \mathscr{P}$. This completes the proof. ∎

Proof of Theorem 4.1

We define an auxiliary variable e,

$$e = E + x - y, e \in \mathscr{E}. \tag{4.34}$$

In the next proof, we use e instead of x, y in (4.11) since e and E uniquely determine x and y through the definition of e and the fact that x and y are not non-zero at the same time. To prove the theorem, we show in the following that the thresholds $\xi_1(E,p)$ and $\xi_2(E,p)$ are the target e that would minimize the function $\tilde{G}(E,p,x,y)$ in the case when $x \geq 0, y = 0$ and the case when $x = 0, y \geq 0$, respectively. When $x \geq 0, y = 0$, the space of e, denoted by $\mathscr{H}_1$, is given by

$$\mathscr{H}_1 = \{e | e \in \mathscr{E}, 0 \leq e - E \leq u_x\} \tag{4.35}$$

and $\tilde{G}(E,p,x,y)$ is computed by

$$\tilde{G}(E,p,x,y) = f((e-E)/\eta + p) + \lambda \mathbf{E}_{q \in \mathscr{P}}[G(e,q)] \; \forall e \in \mathscr{H}_1. \tag{4.36}$$

When $x = 0, y \geq 0$, the space of e, denoted by $\mathscr{H}_2$, is given by

$$\mathscr{H}_2 = \{e | e \in \mathscr{E}, 0 \leq E - e \leq u_y\} \tag{4.37}$$

and $\tilde{G}(E,p,x,y)$ is computed by

$$\tilde{G}(E,p,x,y) = f(\eta(e-E) + p) + \lambda \mathbf{E}_{q \in \mathscr{P}}[G(e,q)] \; \forall e \in \mathscr{H}_2. \tag{4.38}$$

For notation brevity, we define function $\hat{G}_1(E,p,e)$ and $\hat{G}_2(E,p,e)$ as

$$\hat{G}_1(E,p,e) = f((e-E)/\eta + p) + \lambda \mathbf{E}_{q \in \mathscr{P}}[G(e,q)] \tag{4.39a}$$

$$\hat{G}_2(E,p,e) = f(\eta(e-E) + p) + \lambda \mathbf{E}_{q \in \mathscr{P}}[G(e,q)], \tag{4.39b}$$

for all $e \in \mathscr{R}$ respectively. Thus, the cost-to-go for $e \in \mathscr{H}_1$, denoted by $G_1(E,p)$, and the cost-to-go for $e \in \mathscr{H}_2$, denoted by $G_2(E,p)$, are computed by

$$G_1(E,p) = \min_{e \in \mathscr{H}_1} \hat{G}_1(E,p,e), \tag{4.40}$$

$$G_2(E,p) = \min_{e \in \mathscr{H}_2} \hat{G}_2(E,p,e), \tag{4.41}$$

respectively. Specially, when $x = 0, y = 0$, we have $e = E$, and the cost-to-go, denoted by $G_3(E,p)$, is computed by

$$G_3(E,p) = \min_{e=E} f(p) + \lambda \mathbf{E}_{q \in \mathscr{P}}[G(e,q)]. \tag{4.42}$$

The optimal charging policy to (4.11) can be obtained by solving (4.40)–(4.42) with variable e in different regions, and select the one that yields the minimum optimal values of (4.40)–(4.42), i.e.,

$$G^*(E,p) = \min\{G_1(E,p), G_2(E,p), G_3(E,p)\}. \tag{4.43}$$

Intuitively, the optimal solutions to (4.40), (4.41), (4.42) are the optimal solution if the battery is under charging mode, discharging mode and idle mode, respectively. By Lemma 4.1, both $\hat{G}_1(E,p,e)$ and $\hat{G}_2(E,p,e)$ are convex in e for all E,p. We denote the minimizer of $\hat{G}_1(E,p,e)$ and $\hat{G}_2(E,p,e)$ by $\xi_1(E,p)$ and $\xi_2(E,p)$ respectively. By definition, $\xi_1(E,p)$ satisfies the condition that the first order derivative of $\hat{G}_1(E,p,e)$ at e is equal to 0, i.e.,

$$f'\left((\xi_1(E,p) - E)/\eta + p\right)/\eta + \lambda \left(\mathbf{E}_{q \in \mathscr{P}}[G(\xi_1(E,p),q)]\right)' = 0. \tag{4.44}$$

Similarly, we have

$$\eta f'\left(-\eta(E - \xi_2(E,p)) + p\right) + \lambda \left(\mathbf{E}_{q \in \mathscr{P}}[G(\xi_2(E,p),q)]\right)' = 0. \tag{4.45}$$

Now we show that

$$\xi_1(E,p) \le \xi_2(E,p) \forall E \in \mathscr{E}, p \in \mathscr{P} \tag{4.46}$$

by the contradiction. Assume that $\xi_1(E,p) > \xi_2(E,p)$. Then,

$$(\xi_1(E,p) - E)/\eta + p > -\eta(E - \xi_2(E,p)) + p. \tag{4.47}$$

Since $f'(.)$ is a non-decreasing function, then based on (4.47),

$$f'\left((\xi_1(E,p) - E)/\eta + p\right)/\eta \ge \eta f'\left(-\eta(E - \xi_2(E,p)) + p\right). \tag{4.48}$$

Combining (4.44), (4.45), (4.48), we have

$$\left(\mathbf{E}_{q \in \mathscr{P}}[G(\xi_1(E,p),q)]\right)' \le \left(\mathbf{E}_{q \in \mathscr{P}}[G(\xi_2(E,p),q)]\right)'. \tag{4.49}$$

Since $\mathbf{E}_{q \in \mathscr{P}}[G(e,q)]$ is a convex function of e, as shown in Lemma 4.1, then, $\left(\mathbf{E}_{q \in \mathscr{P}}[G(e,q)]\right)'$ is a non-decreasing function of e. Thus, there is $\xi_1(E,p) \le \xi_2(E,p)$, which leads to the contradiction with the assumption that $\xi_1(E,p) > \xi_2(E,p)$. Next, we divide all possible relationships between $\xi_1(E,p), \xi_2(E,p), E, 0, \zeta$ into the following five cases to discuss the optimal charging policy.

Case 1: If $0 \le E \le \zeta \le \xi_1(E,p)$, we calculate the optimal e under three battery modes respectively. In other words, we calculate the optimal solution to (4.40), (4.41), (4.42) respectively. When the battery is under idle mode, i.e., $e = E$, the optimal solution to (4.42) is trivial to be E, and the optimal value is $G_3(E,p) = f(p) + \lambda \mathbf{E}_{q \in \mathscr{P}}[G_t(E,q)]$. When the battery is under charging mode, i.e., $e \in \mathscr{H}_1$, $\hat{G}_1(E,p,e)$ decreases as e increases within the domain $(E, \zeta]$. Then, the optimal solution to (4.40) is given by $e = \min\{\zeta, E + u_x\}$ and the optimal value is given by $G_1(E,p) = \hat{G}_1(E,p,\min\{\zeta, E + u_x\})$. When the battery is under discharging mode, i.e., $e \in \mathscr{H}_2$, $\hat{G}_2(E,p,e)$ increases as e decreases within the domain $[0, E]$. Then, the optimal solution to (4.41) is $e = E$ and the optimal value is given by $G_2(E,p) = \hat{G}_2(E,p,E)$. Thus, based on the fact that the $\hat{G}_1(E,p,e)$ decreases within the domain $[E, \zeta]$, we have the following inequality.

$$G_1(E,p) = \hat{G}_1(E,p,\min\{\zeta, E + u_x\}) \tag{4.50a}$$

$$\le \hat{G}_1(E,p,E) \tag{4.50b}$$

$$= f(p) + \lambda \mathbf{E}_{q \in \mathscr{P}}[G_t(E,q)] \tag{4.50c}$$

$$= G_3(E,p). \tag{4.50d}$$

On the other hand,

$$G_2(E,p) = \hat{G}_2(E,p,E) = f(p) + \lambda \mathbf{E}_{q \in \mathscr{P}}[G_t(E,q)] = G_3(E,p). \tag{4.51}$$

Therefore, the optimal cost-to-go is given by

$$G^*(E,p) = G_1(E,p) = \hat{G}_1(E,p,\min\{\zeta, E + u_x\}) \tag{4.52}$$

and the optimal charging policy is given by

$$x^* = \min\{\zeta - E, u_x\}, y^* = 0. \tag{4.53}$$

Case 2: If $0 \le E \le \xi_1(E,p) \le \zeta$, we calculate the optimal value under three battery modes respectively. When the charging is under idle or discharge mode, the results are the same as Case 1. When the battery is under charging mode, i.e., $e \in \mathcal{H}_1$, $\hat{G}_1(E,p,e)$ decreases as e increases within the domain $(E,\xi_1(E,p)]$. Thus, the optimal solution to (4.40) is given by $e = \min\{\xi_1(E,p),E+u_x\}$ and the optimal value is given by $G_1(E,p) = \hat{G}_1(E,p,\min\{\xi_1(E,p),E+u_x\})$. Since $\hat{G}_1(E,p,e)$ decreases within the domain $[E,\xi_1(E,p)]$,

$$G_1(E,p) = \hat{G}_1(E,p,\min\{\xi_1(E,p),E+u_x\}) \tag{4.54a}$$

$$\leq \hat{G}_1(E,p,E) \tag{4.54b}$$

$$= f(p) + \lambda \mathbf{E}_{q\in\mathscr{P}}[G_t(E,q)] \tag{4.54c}$$

$$= G_3(E,p). \tag{4.54d}$$

Thus, the optimal cost-to-go is given by

$$G^*(E,p) = G_1(E,p) = \hat{G}_1(E,p,\min\{\xi_1(E,p),E+u_x\}), \tag{4.55}$$

and the optimal charging policy is given by

$$x^* = \min\{\xi_1(E,p)-E,u_x\}, y^* = 0. \tag{4.56}$$

Case 3: If $\xi_2(E,p) \leq 0 \leq E \leq \zeta$, we still solve (4.40), (4.41), (4.42) respectively. When the battery is under idle mode, i.e., $e = E$, the optimal solution to (4.42) is $e = E$, and the optimal value is $G_3(E,p) = f(p) + \lambda \mathbf{E}_{q\in\mathscr{P}}[G_t(E,q)]$. When the battery is under discharging mode, i.e., $e \in \mathcal{H}_2$, $\hat{G}_2(E,p,e)$ decreases as e decreases within the domain $(0,E]$. Then, the optimal solution to (4.41) is given by $e = \max\{0,E-u_y\}$ and the optimal value is given by $G_2(E,p) = \hat{G}_2(E,p,\max\{0,E-u_y\})$. When the battery is under charging mode, i.e., $e \in \mathcal{H}_1$, $\hat{G}_1(E,p,e)$ increases as e increases within the domain $[E,\zeta]$ Then, the optimal solution to (4.40) is $e = E$ and the optimal value is given by $G_1(E,p) = \hat{G}_1(E,p,E)$. By the fact that $\hat{G}_2(E,p,e)$ increases within the domain $[0,E]$, we have the following inequality.

$$G_2(E,p) = \hat{G}_2(E,p,\max\{0,E-u_y\}) \tag{4.57a}$$

$$\leq \hat{G}_2(E,p,E) \tag{4.57b}$$

$$= f(p) + \lambda \mathbf{E}_{q\in\mathscr{P}}[G_t(E,q)] \tag{4.57c}$$

$$= G_3(E,p). \tag{4.57d}$$

On the other hand,

$$G_1(E,p) = \hat{G}_1(E,p,E) = f(p) + \lambda \mathbf{E}_{q\in\mathscr{P}}[G_t(E,q)] = G_3(E,p). \tag{4.58}$$

Thus, the optimal cost-to-go is given by

$$G^*(E,p) = G_2(E,p) = \hat{G}_2(E,p,\max\{0,E-u_y\}), \tag{4.59}$$

and the optimal charging policy is given by

$$x^* = 0, y^* = \min\{E,u_y\}. \tag{4.60}$$

Case 4: If $0 \leq \xi_2(E,p) \leq E \leq \zeta$, similar to Case 3, the optimal cost-to-go is given by

$$G^*(E,p) = G_2(E,p) = \hat{G}_2(E,p,\max\{\xi_2(E,p),E-u_y\}), \tag{4.61}$$

and the optimal charging policy is given by

$$x^* = 0, y^* = \min\{E-\xi_2(E,p),u_y\}. \tag{4.62}$$

Case 5: If $\xi_1(E,p) < E < \xi_2(E,p)$, we can observe that under the charging mode $\hat{G}_1(E,p,e)$ increases as e increases within $(E,\xi_2(E,p)]$ and under the discharging mode $\hat{G}_2(E,p,e)$ increases as e decreases within $[\xi_1(E,p),E)$. That is, there is no incentive to charge or discharge. i.e.,

$$G_1(E,p) = \hat{G}_1(E,p,E) = G_3(E,p) \tag{4.63}$$

and

$$G_2(E,p) = \hat{G}_2(E,p,E) = G_3(E,p). \tag{4.64}$$

Therefore, the optimal cost-to-go is given by

$$G^*(E,p) = G_1(E,p) = G_2(E,p) = G_3(E,p), \tag{4.65}$$

and the optimal charging policy is given by

$$x^* = 0, y^* = 0. \tag{4.66}$$

Combining the results of Case 1 and Case 2, Case 3 and Case 4, we can get the first two equalities in (4.14) respectively, and Case 5 verifies the third equality in (4.14). This completes the proof. ∎

Proof of Proposition 4.2

In order to show the proposition, it is sufficient to show that the function $\tilde{G}(E,p,x,y)$ is an increasing function of x and decreasing function of y for all $x,y \in \mathscr{S}(E), E \in \mathscr{E}, p \in \mathscr{P}, y \in \mathscr{S}(E)$. That is, for all $x_1 \leq x_2, x_1, x_2 \in \mathscr{S}(E), E \in \mathscr{E}, p \in \mathscr{P}, y \in \mathscr{S}(E)$, we have

$$\tilde{G}(E,p,x_1,y) \leq \tilde{G}(E,p,x_2,y), \tag{4.67}$$

and for all $y_1 \geq y_2, y_1, y_2 \in \mathscr{S}(E), E \in \mathscr{E}, p \in \mathscr{P}, x \in \mathscr{S}(E)$, we have

$$\tilde{G}(E,p,x,y_1) \leq \tilde{G}(E,p,x,y_2). \tag{4.68}$$

Now we show that (4.67) holds for all $x_1 \leq x_2, x_1, x_2 \in \mathscr{S}(E), E \in \mathscr{E}, p \in \mathscr{P}, y \in \mathscr{S}(E)$. Equation (4.67) is equivalent to the following inequality.

$$\begin{aligned} \mathbf{E}_{q \in \mathscr{P}}[G(E+x_1-y,q)-G(E+x_2-y,q)] \leq \\ (x_2-x_1)\left((x_1+x_2)/\eta - 2\eta y + 2p\right)/(\eta\lambda). \end{aligned} \tag{4.69}$$

Since $x_1, x_2, y \in \mathscr{S}(E)$, then,

$$(x_1+x_2)/\eta - 2\eta y + 2p \geq 2p - 2\eta u_y \geq 2p_1 - 2\eta u_y. \tag{4.70}$$

Combining (4.15) and (4.70), we have

$$(x_2-x_1)\left((x_1+x_2)/\eta - 2\eta y + 2p\right)/(\eta\lambda) \geq (x_2-x_1)(2p_2+2u_x/\eta). \tag{4.71}$$

Thus, to prove (4.69) holds, it suffices to show that the inequality

$$G(E+x_1-y,q)-G(E+x_2-y,q) \leq (x_2-x_1)(2p_2+2u_x/\eta) \tag{4.72}$$

holds for all $q \in \mathscr{P}$. Next, we show that (4.72) holds by the mathematical induction. Define $G^k(E,q)$ as the cost-to-go value at kth value iteration for all $E \in \mathscr{E}, q \in \mathscr{P}$ in Algorithm 3. When $k=0$,

$$\begin{aligned} G^0(E+x_1-y,q) - G^0(E+x_2-y,q) \\ =0 \leq (x_2-x_1)(2p_2+2u_x/\eta). \end{aligned} \tag{4.73}$$

Now assume that $\forall q \in \mathscr{P}, k=0,1,2,\cdots,$

$$G^k(E+x_1-y,q)-G^k(E+x_2-y,q) \leq (x_2-x_1)(2p_2+2u_x/\eta). \tag{4.74}$$

Let x_1^{k+1}, y_1^{k+1} and x_2^{k+1}, y_2^{k+1} be the solutions that yield the cost-to-go value at $(k+1)$th value iteration, i.e., $G^{k+1}(E+x_1-y,q)$ and $G^{k+1}(E+x_2-y,q)$ respectively. Then,

$$G^{k+1}(E+x_1-y,q) - G^{k+1}(E+x_2-y,q)$$

$$= \left(\frac{x_1^{k+1}}{\eta} - \eta y_1^{k+1} + q\right)^2 - \left(\frac{x_2^{k+1}}{\eta} - \eta y_2^{k+1} + q\right)^2$$

$$+ \lambda \mathbf{E}_{q' \in \mathcal{P}}[G^k(E+x_1-y+x_1^{k+1}-y_1^{k+1},q')]$$

$$- \lambda \mathbf{E}_{q' \in \mathcal{P}}[G^k(E+x_2-y+x_2^{k+1}-y_2^{k+1},q')] \tag{4.75a}$$

$$= \left(\frac{x_1^{k+1}+x_2^{k+1}}{\eta} - \eta(y_1^{k+1}+y_2^{k+1}) + 2q\right)\left(\frac{x_1^{k+1}-x_2^{k+1}}{\eta}\right)$$

$$- \eta(y_1^{k+1}-y_2^{k+1})\right) + \lambda \mathbf{E}_{q' \in \mathcal{P}}[G^k(E+x_1-y+x_1^{k+1}$$

$$- y_1^{k+1},q') - G^k(E+x_2-y+x_2^{k+1}-y_2^{k+1},q')] \tag{4.75b}$$

$$\leq \left(\frac{x_1^{k+1}+x_2^{k+1}}{\eta} - \eta(y_1^{k+1}+y_2^{k+1}) + 2q\right)\left(\frac{x_1^{k+1}-x_2^{k+1}}{\eta}\right)$$

$$- \eta(y_1^{k+1}-y_2^{k+1})\right) + \lambda(x_2^{k+1}-x_1^{k+1}-\eta(y_2^{k+1}-y_1^{k+1})$$

$$+ x_2 - x_1)(2p_2 + 2u_x/\eta), \tag{4.75c}$$

where the last inequality holds based on (4.74). Since

$$x_1^{k+1} \leq u_x, x_2^{k+1} \leq u_x, y_1^{k+1} \geq 0, y_2^{k+1} \geq 0, \tag{4.76}$$

then

$$\frac{x_1^{k+1}+x_2^{k+1}}{\eta} - \eta(y_1^{k+1}+y_2^{k+1}) + 2q \leq 2u_x/\eta + 2q \tag{4.77a}$$

$$\leq 2u_x/\eta + 2p_2, \tag{4.77b}$$

where the last inequality holds because $q \leq p_2$. By substituting (4.77) into (4.75), we have

$$G^{k+1}(E+x_1-y,q) - G^{k+1}(E+x_2-y,q)$$

$$\leq (2u_x/\eta + 2p_2)\left(\frac{x_1^{k+1}-x_2^{k+1}}{\eta} - \eta(y_1^{k+1}-y_2^{k+1})\right)$$

$$+ \lambda \left(\frac{x_2^{k+1} - x_1^{k+1}}{\eta} - \eta (y_2^{k+1} - y_1^{k+1}) + x_2 - x_1 \right) (2p_2$$

$$+ 2u_x/\eta) \tag{4.78a}$$

$$\leq (x_2 - x_1)(2u_x/\eta + 2p_2), \tag{4.78b}$$

where the last inequality holds because $0 < \eta \leq 1, \lambda \leq 1$. Since

$$\lim_{k \to \infty} G^k(E + x_1 - y, q) - G^k(E + x_2 - y, q)$$
$$= G(E + x_1 - y, q) - G(E + x_2 - y, q) \tag{4.79}$$

holds for all $q \in \mathscr{P}$, it is sufficient to show that (4.72) holds. Similarly, we can show that (4.68) holds for any $y_1 \geq y_2, y_1, y_2 \in \mathscr{S}(E)$. This completes the proof. ∎

Proof of Proposition 4.2

For any $\zeta' \geq 0$, define $x^*(\zeta'), y^*(\zeta')$ as optimal solution to (4.11) with $\zeta = \zeta'$. Likewise, we define $x^*(\zeta''), y^*(\zeta'')$ for any $\zeta'' \geq 0$. Now let

$$\zeta''' = \alpha \zeta' + (1 - \alpha)\zeta'', \forall \alpha \in [0, 1]. \tag{4.80}$$

Then, there must exist a group of $x(\zeta'''), y^*(\zeta''')$ such that

$$x(\zeta''') = \alpha x^*(\zeta') + (1 - \alpha)x^*(\zeta''), y(\zeta''') = \alpha y^*(\zeta') + (1 - \alpha)y^*(\zeta''). \tag{4.81}$$

Note that $x(\zeta'''), y(\zeta''')$ still satisfies $x(\zeta'''), y(\zeta''') \in \mathscr{S}(E)$ due to the linearity of (4.8). Based on (4.81), we have

$$\alpha(x_t^*(\zeta')/\eta - \eta y_t^*(\zeta') + p_t) + (1 - \alpha)(x_t^*(\zeta'')/\eta - \eta y_t^*(\zeta'') + p_t)$$
$$= x_t(\zeta''')/\eta - \eta y_t(\zeta''') + p_t, \alpha \in [0, 1], \tag{4.82}$$

and

$$\alpha(E + x^*(\zeta') - y^*(\zeta')) + (1 - \alpha)(E + x^*(\zeta'') - y^*(\zeta''))$$
$$= E + x(\zeta''') - y^*(\zeta'''), \forall \alpha \in [0, 1]. \tag{4.83}$$

Then, we have

$$f(x(\zeta''')/\eta - \eta y(\zeta''') + p) \leq \alpha f(x^*(\zeta')/\eta - \eta y^*(\zeta') + p_t)$$
$$+ (1 - \alpha)f(x^*(\zeta'')/\eta - \eta y^*(\zeta'') + p_t) \tag{4.84}$$

and

$$
\begin{aligned}
\mathbf{E}_{q\in\mathscr{P}}[G^*(E+x(\zeta''')-y(\zeta'''),q)] &\leq \alpha\mathbf{E}_{q\in\mathscr{P}}[G^*(E+x^*(\zeta') \\
-y^*(\zeta'),q)]&+(1-\alpha)\mathbf{E}_{q\in\mathscr{P}}[G^*(E+x^*(\zeta'')-y^*(\zeta''),q)]
\end{aligned}
\tag{4.85}
$$

for all $\alpha \in [0,1]$, where inequality (4.84) and (4.85) hold due to the convexity of $f(x(\zeta''')/\eta - \eta y(\zeta''') + p)$ and $\mathbf{E}_{q\in\mathscr{P}}[G^*(E+x(\zeta''')-y(\zeta'''),q)]$, respectively. Thus, Combining (4.84) and (4.85), we have

$$
\begin{aligned}
&\tilde{G}(E,p,x(\zeta'''),y(\zeta''')) \\
&\leq \alpha\tilde{G}(E,p,x^*(\zeta'),y^*(\zeta'))+(1-\alpha)\tilde{G}(E,p,x^*(\zeta''),y^*(\zeta''))
\end{aligned}
\tag{4.86}
$$

for all $\alpha \in [0,1]$. Let $x^*(\zeta'''),y^*(\zeta''')$ be the optimal solution to (4.11) with $\zeta = \zeta'''$. By definition,

$$
\tilde{G}(E,p,x^*(\zeta'''),y^*(\zeta''')) \leq \tilde{G}(E,p,x(\zeta'''),y(\zeta''')).
\tag{4.87}
$$

Based (4.86) and (4.87), we have

$$
\begin{aligned}
&\tilde{G}(E,p,x^*(\zeta'''),y^*(\zeta''')) \\
&\leq \alpha\tilde{G}(E,p,x^*(\zeta'),y^*(\zeta'))+(1-\alpha)\tilde{G}(E,p,x^*(\zeta''),y^*(\zeta''))
\end{aligned}
\tag{4.88}
$$

for all $\alpha \in [0,1]$. Thus, we have established the convexity of $G^*(E,p)$ over $\zeta, \zeta \geq 0$. Next, we show that $G^*(E,p)$ is a decreasing function of ζ. Suppose that $\zeta' \leq \zeta''$. Then, by definition of $x^*(\zeta')$ and $y^*(\zeta')$, $x^*(\zeta')$ and $y^*(\zeta')$ satisfy

$$
0 \leq E+x^*(\zeta')-y^*(\zeta') \leq \zeta' \leq \zeta'',
\tag{4.89}
$$

which indicates that $x^*(\zeta'),y^*(\zeta')$ are the feasible solutions to (4.11) with $\zeta = \zeta''$. Since $x^*(\zeta''),y^*(\zeta'')$ are the optimal solutions to (4.11) with $\zeta = \zeta''$, then, we have

$$
\tilde{G}(E,p,x^*(\zeta'),y^*(\zeta')) \geq \tilde{G}(E,p,x^*(\zeta''),y^*(\zeta'')).
\tag{4.90}
$$

This completes the proof.   ∎

References

1. W. Tang, S. Bi, and Y. J. Zhang, "Online coordinated charging decision algorithm for electric vehicles without future information," *IEEE Trans. on Smart Grid*, vol. 5, no. 6, pp. 2810–2824, 2014.
2. Z. Ma, D. Callaway, and I. Hiskens, "Decentralized Charging Control of Large Populations of Plug-in Electric Vehicles", *IEEE Trans. on Control Systems Technology*, vol.21, no.1, pp. 67–78, 2013.
3. Y. He, B. Venkatesh, and L. Guan, "Optimal scheduling for charging and discharging of electric vehicles," *IEEE Trans. on Smart Grid*, vol.3, no.3, pp. 1095–1105, 2012.
4. W. Tang and Y. J. Zhang, "A model predictive control approach for low-complexity electric vehicle charging scheduling: optimality and scalability," *IEEE Trans. Power Systems*, Jun. 2016.
5. V. Aguirregabiria, "Estimation of dynamic programming models with censored dependent variables," *Investigaciones Economicas*, vol. XXI(2), pp. 167–208, 1997.
6. J. Qin, R. Sevlian, D. Varodayan, and R. Rajagopal, "Optimal electric energy storage operation," *in Proc. of IEEE PES General Meeting*, pp. 1–6, Jul. 2012.
7. P. van de Ven, N. Hegde, L. Massouli'e, and T. Salonidis, "Optimal control of end-user energy storage," *IEEE Trans. on Smart Grid*, vol. 4, no. 2, pp. 789–797, 2013.
8. Y. Xu and L. Tong, "On the operation and value of storage in consumer demand response," *53rd IEEE Conference on Decision and Control*, pp. 205–210, Dec. 2014.
9. Y. Ru, J. Kleissl, and S. Martinez, "Storage size determination for grid-connected photovoltaic systems," *IEEE Trans. on Sustainable Energy*, vol. 4, no. 1, pp. 68–81, 2013.
10. I. Koutsopoulos, V. Hatzi, and L. Tassiulas, "Optimal energy storage control policies for the smart power grid," *Proc. of IEEE International Conference on Smart Grid Communications (SmartGridComm)*, pp. 475–480, Oct. 2011.
11. L. Huang, J. Walrand, and K. Ramchandran, "Optimal demand response with energy storage management," *Proc. of IEEE International Conference on Smart Grid Communications (SmartGridComm)*, pp. 61–66, Nov. 2012.
12. J. Qin, Y. Chow, J. Yang, and R. Rajagopal, "Modeling and Online Control of Generalized Energy Storage Networks," *in Proc. of the 5th International Conference on Future Energy Systems (ACM e-Energy '14). ACM*, pp. 27–38, Jun. 2014.
13. D. P. Bertsekas, *Dynamic Programming and Optimal Control*. Belmont, MA: Athena Scientific, 2007.
14. A. Ipakchi and F. Albuyeh, "Grid of the future," *IEEE Power and Energy Mag.*, vol. 7, no. 2, pp. 52–62, 2009.
15. EPRI, *Handbook of Energy Storage for Transmission or Distribution Applications*, 2002.

Chapter 5
Conclusions and Future Work

5.1 Conclusions

Grid level energy storage systems are a cornerstone of future power networks and smart grid development. Being aware of challenges in different type of energy storage systems, this book focuses on proposing online charging control scheme to minimize the total energy cost and flatten the total power flow exchanged with the main grid. The main results of the book are summarized as follows.

Chapter 2 investigates the PEV charging problem in a community, whose power consumption consists of the load of a PEV charging station and the other inelastic base load besides the PEV charging consumption. By controlling the charging rates of PEVs, we aim to minimize total cost on electricity bill paid by the charging station. we propose an Online cooRdinated CHARging Decision (ORCHARD) algorithm, which minimizes the energy cost without knowing the future information. Through rigorous proof, we show that ORCHARD is strictly feasible in the sense that it guarantees to fulfill all charging demands before due time. Meanwhile, it achieves the best known competitive ratio of 2.39 when the cost function is a quadratic function of the load demand. By exploiting the problem structure, we propose a novel reduced-complexity algorithm to replace the standard convex optimization techniques used in ORCHARD. Besides, we show that this reduced-complexity algorithm output the optimal solution to the offline PEV charging problem.

In Chap. 3, we consider the optimal PEV charging scheduling, which results in a charging load demand that is as flat as possible over time. Specifically, we consider a practical scenario, where the non-causal information about future PEV arrivals is not known in advance, but its statistical information can be estimated. This leads to an "online" charging scheduling problem that is naturally formulated as a finite-horizon dynamic programming with continuous state space and action space. We provide a Model Predictive Control (MPC) based algorithm with computational complexity

© The Author(s) 2017
W. Tang, Y.J.A. Zhang, *Optimal Charging Control of Electric Vehicles in Smart Grids*, SpringerBriefs in Electrical and Computer Engineering, DOI 10.1007/978-3-319-45862-5_5

$O(T^3)$, where T is the total number of time stages. We rigorously prove that the proposed algorithm yields a near-optimal solution that has a bounded performance gap from the optimal solution regardless of the distribution of exogenous random variables. Furthermore, our rigorous analysis shows that the proposed algorithm can be made scalable when the random process describing the arrival of charging demands is first-order periodic.

Chapter 4 is concerning the optimal control of a battery energy storage system (BESS) in a microgrid with renewable energy sources. In particular, the optimal control decisions are made without the non-causal knowledge of future load demand and renewable source output. We aim to minimize a cost function, which is a general convex increasing function of instantaneous power flow from the grid. The increasing convexity of the cost function makes sure that the power flow exchanged with the main grid is as flat over time as possible, and thus minimizes the negative impact of renewable energy integration. Through rigorous analysis, we prove that the optimal BESS operation policy exhibits a threshold structure, which can potentially lead to a simplified control policy. Specifically, we show that the optimal policy degenerates to one that takes a *short-sighted* behavior when the discount factor satisfies certain conditions. Moreover, we show that the optimal cost is a decreasing convex function of the battery capacity, implying that there exists an optimal battery sizing that strikes a balance between the operation cost and the capital investment.

To sum up, we mainly studied the online charging control for two energy storage systems, i.e., MESS and BESS, in different scenarios. We prove that the proposed online charging schemes achieve remarkable improvement on the optimality and scalability of system performance. Our results also show that the proposed online charging schemes can efficiently save the energy cost as well as reducing the fluctuation of the total load output from the main grid.

5.2 Future Work

The work presented in this book offers many possibilities for future extensions. In particular, the following topics are of interest:

1. The PEV charging problem studied in this book assumes that the charging rate of PEVs are non-negative, where the power only transmits from the grid to the energy system stored in the PEV. In other words, we only consider the case of Grid to Vehicle (G2V). With significant penetration of PEVs in the near future, the concept introduced in literatures as Vehicle to Grid (V2G) will be practically possible [1]. The V2G concept eases the integration of renewable energy resources into power system and gives a new force to the inevitable move towards power generation by clean energy resources. Then, it is critical to construct and analyze a general system which includes G2V, V2G and the renewable energy generators. Meanwhile, the control schemes can be designed

for different purposes, such as minimizing the total operation cost, stabilizing the voltage, flattening the total power flow, etc. The analysis of the worst-case performance and average-case performance of the proposed algorithms in this book may be extended to the control of the general system.

2. The major challenge of the online charging algorithm design is the uncertainties from the behavior of EV users. A promising solution is to introduce economic incentive schemes to encourage more users to arrive at the charging station during the off-peak hour of base load consumptions and less during the peak hours, so that the total load demand is flattened over time. Equivalently, pricing method can be used to adjust the EVs' charging demand over time. Besides, the scheduler can also offer financial compensation to those users who are willing to make reservations day-ahead, park the EV for a longer time, or tolerate charging delay after the specified parking time. Through optimizing the pricing schemes, the scheduler maximizes its overall utility, e.g., its profit defined as the revenue minus the operating cost and the cost on offering the incentives. The joint design of pricing scheme and online EV scheduling is also a promising yet challenging topic to investigate, considering the complex correlations between the pricing and the EV user profiles, including arrival rates, parking time and charging demand.

3. Tang and Zhang [2] shows that the accurate knowledge of future data can lead to significant performance improvement of online algorithms. Currently, most studies on online scheduling design assume perfect knowledge of (partial) future data or statistical information. In practice, however, the actual knowledge could be inaccurate, and the data collected could be noisy, incomplete or out-dated. It is therefore important to incorporate the acquisition of data knowledge in the design of online scheduling algorithm. A promising solution is to use online/stochastic learning methods to exploit the random data to assist the decisions of EV scheduling in an iterative manner. In this case, however, the learning algorithm efficiency is of paramount importance, as the EV data size could be enormous and the charging scheduling is a delay-sensitive application.

4. In the BESS control problem, we focus on the scenario where the microgrid includes only one BESS. Another scenario of our interest is that the microgrid includes an aggregator connected a number of BESSs through a distribution network. For the microgrid system with one BESS, the charing/dischaging power flow of the BESS is limited by the physical factors of batteries, i.e., max charging/discharging rate, battery size, etc. In contrast, for the distributed BESS system, the charing/dischaging power flow of each BESS is constrained by not only the physical factors of batteries, but also the voltage of the location and the power loss during the transmission. On the other hand, current research mainly focuses on the centralized optimal control scheme design [3]. In practice, the integration of distributed energy resources would require a decentralized solution to these problems. Then it is an important future direction to incorporate the distributed/decentralized deterministic optimization into the online control scheme.

5. Large-scale deployment of BESSs is considered to be a promising mechanism for fast frequency control in various electric systems. Although BESS can not

compete with the mechanical storage system regarding the capacity of storage, it can play an important part by compensating small deviations from the power balance, among others. The BESS takes instruction from the power grid to provide or absorb a certain amount of power for the purpose of primary frequency control. The profitability of this application is mainly established by comparing frequency control reserve prices on ancillary service markets with realistic installation and maintenance costs of BESS units [4]. To guarantee the full availability for primary frequency control, it requires efficient control algorithms to keep the State of Charge (SoC) in between certain limits.

6. The integration of renewable sources brings both challenges and opportunities to the EV charging scheduling problem. On one hand, EVs as energy storage can be used to reduce the intermittency of renewable sources, absorb the variability of load caused by renewable sources and even as energy carriers to transport energy from remote renewable sources to loads in urgent need of power supply. On the other hand, renewable source could help reduce the fluctuation of base load and energy generation cost, especially for charging stations that own distributed renewable generators. However, the integration of renewable energy introduces another layer of randomness in the system design, such that online algorithms now need to tackle the uncertainties from both the EVs and the renewable sources. Prediction and data mining play even more important role in improving the overall system performance.

References

1. S. S. Hosseini, A. Badri, and M. Parvania, "A survey on mobile energy storage systems (MESS): Applications, challenges and solutions," *Renewable Sustainable Energy Rev. 40*, pp. 161–170, 2014.
2. W. Tang and Y. J. Zhang, "A model predictive control approach for low-complexity electric vehicle charging scheduling: optimality and scalability," accepted by *IEEE Trans. Power Syst.*, Jun. 2016.
3. J. Qin, Y. Chow, J. Yang, and R. Rajagopal, "Modeling and Online Control of Generalized Energy Storage Networks," *in Proc. of the 5th International Conference on Future Energy Systems (ACM e-Energy '14). ACM*, pp. 27–38, Jun. 2014.
4. A. Oudalov, D. Chartouni, and C. Ohler, "Optimizing a Battery Energy Storage System for Primary Frequency Control," *IEEE Trans. on Power System*, vol. 22, no. 3, pp. 1259–1266, 2007.

Printed in the United States
By Bookmasters